Instructive Wound Dressings Silk Fibroin and Paramylon Revolution

Abigail Cowen

First Printing, 2024

TABLE OF CONTENTS

CHAPTER NO. TITLE PAGE NO.

CHAPTER NO. TITLE PAGE NO.

CHAPTER NO. TITLE PAGE NO.

CHAPTER NO. TITLE PAGE NO.

CHAPTER NO. TITLE PAGE NO.

CHAPTER NO. TITLE PAGE NO.

ABSTRACT

Chronic wounds are one among the dangerous issues that are predicted to affect 2-6 crore people around the world by 2026 due to the endless rising of obesity, aging, diabetes and change in lifestyle. These wounds are more than just a cost burden, as they create immobility and worthlessness of life due to the disability to perform daily tasks. Factors responsible for the chronicity in wounds are continuous trauma, debilitating illness, increased bacterial protein, ischemia, excessive proteases, destruction of extracellular matrix (ECM) and growth factors that are mandatory for healing. Thus, to positively regulate the proteases like Matrix metalloproteases and to activate the growth factors and ECM, Biomimetics is one of the favourable approaches to develop an ECM like dressing to progress the wound healing process in chronic wounds. Due to the increasing interest and the ECM framework, Protein-Polysaccharide complex driven the curiosity in developing a biomimetic and instructive wound dressing usind silk fibroin protein and Paramylon polysaccharide-based complex of different forms have been studied to develop a self-care wound dressing for chronic wounds.

Primarily to study the suitability, silk fibroin and paramylon blended films was prepared in different ratio (100:0, 75:25, 50:50, 75:25 and 0:100). Later three forms of Silk fibroin and paramylon-based materials was fabricated such as (1) Silk Fibroin based non-woven mat functionalized with 0.25%, 0.5% and 1% paramylon (2) Alum loaded FG 75:25 coated cellulose sponge for exudate wounds and (3) Paramylon coated silk fibroin sponge sheet. The surface topography, molecular confirmation, wettability, porosity, swelling capacity, enzymatic degradation, hemolysis and cytotoxicity studies were studied for all three forms. In which, one from each was considered to study the wound healing activity along with the paramylon release kinetics. It was observed that based on the swelling, degradation, porosity measurement and cytotoxicity studies, the nonwoven mats SBM25 was further studied for knowing its release and wound healing property that resulted in 92-99% of wound closure with 50% release of paramylon release. In the sponge forms, CAP and BPFS was considered based on the cytotoxicity and moderate swelling capacity. Eventhough CAP showed 97-99% of wound closure with about 50% release of paramylon, BPFS sponge showed only 37-49% wound closure with low release profile of paramylon below 19%. Thus, SBM25 nonwoven mat and CAP sponge was suggested for the future studies to be optimized and used as bandage for low depth wound and in wounds with more exudates respectively.

LIST OF TABLES

LIST OF FIGURES

FIGURE NO. TITLE PAGE NO.

FIGURE NO. TITLE PAGE NO.

1. Scratch wound assay using L929 fibroblasts incubated with SBM25 nonwoven mat

(25, 50, 75 & µg) at 0, 4, 18, 24 hours 65

1. Percentage of wound closure at 0, 4, 18, 24 hours. Data represent mean±standard error of the

mean(n=3) 65

FIGURE NO. TITLE PAGE NO.

FIGURE NO. TITLE PAGE NO.

1. Intrinsic coagulation factors activation measurement curve. Data represent

mean±standard error of the mean(n=3) 104

LIST OF SYMBOLS AND ABBREVIATIONS

μL MicroLitre

VEGF Vascular Endothelial Growth Factor

V Volume

Upa Urokinase Plasminogen Activator

TNF a Tumor Necrosis Factor - Alpha

TLR Toll like Receptor

TIMPs Tissue Inhibitors of Metalloproteinases

TGF � Transforming Growth Factor - Beta

SPG Schizophyllan

SM P Non enzymatic degradation with PBS on Silk Fibroin Nonwoven Mat

SM L Enzymatic degradation with Lysozyme on Silk Fibroin Nonwoven Mat

SM Silk Fibroin Nonwoven Mat

SI Swelling Index

SFS Silk Fibroin Sponge

SF Silk Fibroin

SEM Scanning Electron Microscopy

SBM5 P Non enzymatic degradation with PBS on SF Nonwoven Mat Functionalized with 0.5% BG

SBM5 L Enzymatic degradation with Lysozyme on SF Nonwoven Mat Functionalized with 0.5% BG

SBM50 SF Nonwoven Mat Functionalized with 0.5% BG

SBM5 SF Nonwoven Mat Functionalized with 0.5% BG

SBM25 P Non enzymatic degradation with PBS on SF Nonwoven Mat Functionalized with 0.25% BG

SBM25 L Enzymatic degradation with Lysozyme on SF Nonwoven

	Mat Functionalized with 0.25% BG
SBM25	SF Nonwoven Mat Functionalized with 0.25% BG
SBM1 P	Non enzymatic degradation with PBS on SF Nonwoven Mat Functionalized with 1% BG
SBM1 L	Enzymatic degradation with Lysozyme on SF Nonwoven Mat Functionalized with 1% BG
SBM1	SF Nonwoven Mat Functionalized with 1% BG
RT	Room Temperature
Rpm	Rotation Per Minute
ROS	Reactive Oxygen Species
RGD	Arg Gly Asp
PVA	Polyvinyl Alcohol
PMN	Polymorphonuclear Leukocytes
pH	potential of Hydrogen
PFS	Potash Alum Coated Fibroin Sponge
PDGF	Platelet Derived Growth Factor
PC	Positive control
PBS	Phosphate Buffer Solution
P	1% Potash Alum Solution
Nm	Nanometer
NHEK	Normal Human Epidermal Keratinocytes
NHDF	Normal Human Dermal Fibroblasts
NC	Negative control
NaCl	Sodium Chloride
M<l	Macrophage
MTT	Dimethylthiazolyl Diphenyltetrazolium Bromide
MMP	Matrix metalloproteases
Ml	MilliLitre
Min	Minute

Mg	Milligram
M	Molar
LiBr	Lithium Bromide
kDA	kilo Dalton
IR	Infrared
IL	Interleukin
Hr	Hour
HIV	Human Immunodeficiency Virus
HEK cells	Human Embryonic Kidney Cells
HDFa	Adult Human Dermal Fibroblast
HDF	Human Dermal Fibroblasts
HA	Hyaluronic Acid
FTIR	Fourier Transform Infrared
FGF	Fibroblast Growth Factor
FG 75:25	Silk Fibroin/Paramylon Blend Ratio 75:25 Respectively
FG 50:50	Silk Fibroin/Paramylon Blend Ratio 50:50 Respectively
FG 25:75	Silk Fibroin/Paramylon Blend Ratio 25:75 Respectively
FBS	Fetal Bovine Serum
FB	Fibroblast
FA	Formic Acid
EPC	Endothelial Progenitor Cells
EDX	Energy Dispersive X Ray Analysis
ECM	Extracellular Matrix
DMSO	Dimethylsulfoxide
DMEM	Dulbecco's Modified Eagle Medium
DME	Dulbecco's Modified Eagle
DC	Dendritic Cells
CXCL8	Interleukin 8
CV	Cell Viability

CR	Complement Receptor
CP	Alum Coated Cellulose Sponge
Cm	Centimeter
CAP	Alum Loaded FG 75:25 Blends Coated Cellulose Sponge
$CaCl_2$	Calcium Chloride
CA	FG 75:25 Blends Coated Cellulose Sponge
C	Cellulose Sponge
BPFS	Potash Alum and Paramylon Coated Fibroin Sponge
BG	Beta glucan derived from *Euglene gracilis* or Paramylon
BFS	Paramylon Coated Fibroin Sponge
BALB	Bagg Albino Mouse
AP	FG 75:25 blends with 1% Potash Alum in 80:20 Ratio
AgNPs	Silver Nanoparticles
ACD	Anticoagulant Citrate Dextrose
A	FG 75:25 blends of SF and BG Solution
0C	Degree Celcius
p	Density

1

CHAPTER 1

INTRODUCTION TO CHRONIC WOUNDS AND THE ROLE OF EXTRACELLULAR MATRIX IN WOUND HEALING

This chapter emphasize the theory of the chronic wounds and its complications along with the importance of extracellular matrix in wound healing process. A brief review relevant to the present work is also presented.

1.1 INTRODUCTION

The skin, the largest organ in the body with an average surface area of 1.6-2 m2 and composed of about 15% of the total body weight of an average adult. It possesses a sequence of functions such as acting as an interface with the environment and protects the body against pathogens, controls weight loss, regulates body temperature and prevents dehydration (Ojeh *et al.* 2015). It is also a water-resistant barrier for protecting essential nutrients and for absorbing oxygen required from the outer most layer of cells. The Skin is composed of 3 layers namely epidermis, dermis and hypodermis. The epidermis is the outer-most layer that act as a barrier to protect the skin from an infection. The second layer is called the dermis and consists of connective tissues which comforts the body from stress and strain. The inner-most layer is the fatty subcutaneous hypodermis which consists of larger blood vessels and nerves that insulates the body and absorbs shock (Paul & Sharma 2015).

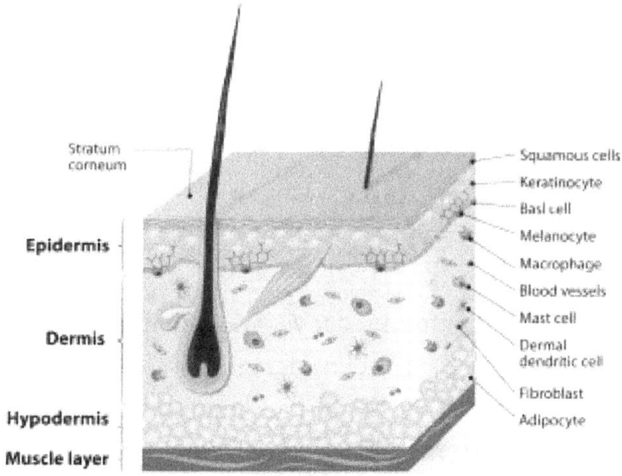

(Source: Johnson 2018)

Figure 1.1 Anatomy of Skin

A Wound is defined as disruption to the normal anatomical structure of the skin that triggers an organized complex cascade of biochemical and cellular events that helps in wound closure (Robson *et al.* 2001). The wounds are generally classified as acute wound and chronic wound with additional subclasses. Acute wounds are surgical incisions and traumatic injuries such as laceration, abrasions, avulsion, penetrations or bites and burn injuries. At the time of injury, all wounds are considered as acute regardless of cause. Wound closure and the establishment of a functional result within acceptable amount of time (8-12 weeks) characterize acute wound (Lazarus *et al.* 1994). The wounds are considered as chronic if the healing time exceeds or if it do not follow the pattern of normal healing (Whitney 2005). The chronic wounds are also an effect of disease, such as venous or arterial vascular insufficiency, cancer, pressure necrosis and diabetes which requires longer healing time and often fail to reach a normal healthy state (Genchi & Ciofani 2019). Therefore, chronic acts as a significant socioeconomic burden to patients and health care systems worldwide, in terms of cost of the treatment and waste production (Sen *et al.* 2009).

1.1.1 Physiology of Wound Healing

In general, Wound Healing is an active and complex process involving series of synchronized events that include coagulation, acute inflammatory response to the initial injury, regeneration, migration and proliferation of connective tissue and parenchyma cells as well as the synthesis of extracellular matrix protein, remodelling of new parenchyma and connective tissue followed by collagen deposition and finally remodelling (Velnar *et al.* 2009). For didactic basis, the wound healing process can be divided into four but synchronizing phases such as hemostasis phase, inflammatory phase, proliferation and maturation phase.

1.1.1.1 Phase 1: Hemostasis

Hemostasis, the first step in wound healing process that begins at the onset of injury and its main objective is to stop bleeding. Upon injury, the blood vessels are damaged and the blood cells move from the capillaries to the nearby tissues (Bryant & Doughty 2000). The initial step is the constriction of blood vessels to reduce the amount of blood loss which is called vascular spasm that is more effective in small blood vessels. It is then triggered by the chemicals released by the endothelial cells and platelets, and as a reflex action initiated by the pain receptors. The platelets in the exposed blood are activated by the exposed collagen fibres and form a gel, a temporary plug for constricting blood vessels to stop bleeding. This temporary plug also acts as a provisional ECM for cell migration. The clotting cascade finally forms fibrin, a mesh-like structure that acts as a framework to trap cellular elements of the blood and reinforces the platelet plug

(Gailit & Clark 1994). The third step of haemostasis is the blood coagulation.

4

The platelet plug secretes wound-healing mediators such as Platelet derived growth factor (PDGF) which activates macrophages and fibroblasts.

In some large injuries, natural hemostasis is difficult and hence applying hemostatic agents topically or direct pressure to bleeding wound may stop bleeding (Glick *et al.* 2013). Also some wounds require suturing to control bleeding.

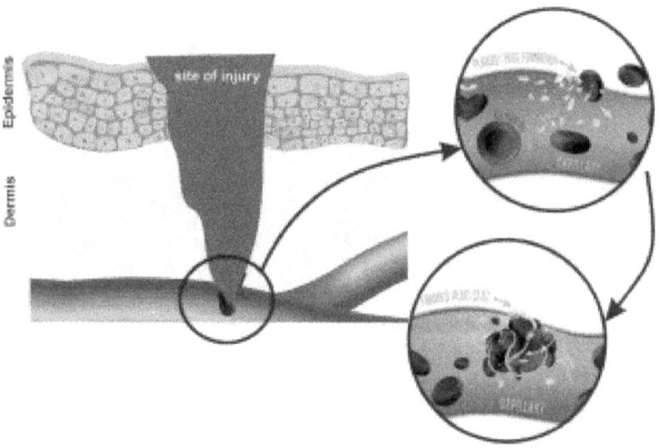

Figure 1.2 Hemostasis

1.1.1.2 Phase 2 : Inflammation

Inflammation is the first overlapping processes that starts the actual process of wound healing. This phase begins immediately after the initial vasoconstriction with the release of prostaglandin and activated complement proteins that causes vasodilation and inflammation (Muire *et al.* 2020). Once haemostasis is achieved, the blood vessels dilate allowing essential components like white blood cells (WBC), growth factors, enzymes and antibodies to reach the site of the wound which in turn increases the exudates level. The body's first defence against microbial invasion starts at this phase.

The infiltrating leukocytes are the principal cellular component of this stage.

5

The polymorphonuclear leukocytes/neutrophils entrapped and aggregated in the blood clot release more growth factors and cytokine signals, and act as chemoattractant for the cells in this phase. Neutrophils begins the debridement of the tissue and phagocytosis of infectious agents (Eming *et al.* 2007). Excessive leukocyte infiltration prolongs the release of oxygen-free radicals, proteolytic enzymes and proinflammatory cytokines that leads to chronicity in wounds. Therefore, a dynamic balance between changes in neutrophil availability and their recruitment to the wound is the main reason for normal wound healing. Neutrophil influx, after a wound occurrence, increased more rapidly in the initial 12 h and reached a maximum value between day 1 and day 2 and sustained up to day 3 and decreased rapid at day 5. Neutrophil accumulation in the wound was 6 times greater than that is present in the entire blood circulation volume on day 2 which returns to normal on wound closure by day

9. Macrophages secrete an angiogenesis factor which stimulates the formation of new blood vessels. These macrophages along with leukocytes serve as phagocytes to clear debris by engulfing and destroying them which is important in the early phase of healing. They release elastases and collagenases for the breakdown of injured tissue and also release cytokines (Kim *et al.* 2008). Platelet-derived growth factor is released by the macrophages which stimulates the chemotaxis and proliferation of fibroblasts and smooth muscle cells. The Macrophages also secrete substances that attract endothelial cells to the wound and stimulate their proliferation to promote angiogenesis. T-lymphocytes that migrate into the wound secrete a basic fibroblast growth factor and heparin-binding epidermal growth factor, which interact with the processes and promote wound healing (Clark & Henson 1996).

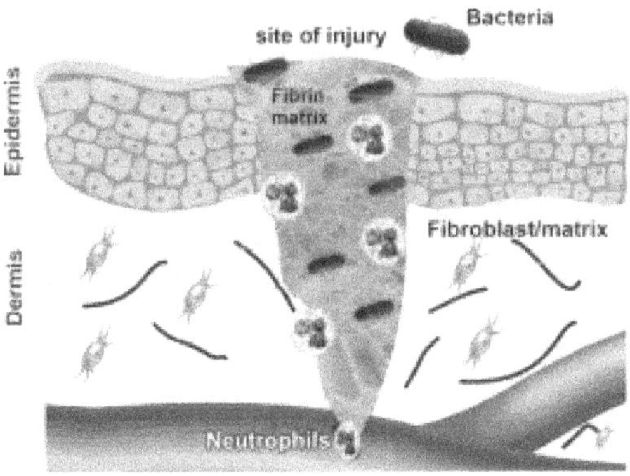

Figure 1.3 Inflammatory Phase

1.1.1.3 Phase 3 : Proliferation

The proliferation phase occurs 3-5 days following injury which is primarily the formation of granulation tissue and may last up to 3 weeks. Granulation tissue consists of a combination of cellular elements, fibroblasts and inflammatory cells, along with new capillaries embedded in a loose extra cellular matrix of collagen, fibronectin and hyaluronic acid (Rajendran 2009). The three stages of the proliferative phase can be termed as granulation, contraction and epithelialization. Angiogenesis induces the formation of new vessels; oxygenated blood reaches the wound bed and the wound becomes less hypoxic and nutrient deficient. The macrophages recruit a new cell type, i.e., the fibroblast, which lays down a network of collagen fibers surrounding the neovasculature of the wound. Fibroblasts proliferate and produce the matrix proteins fibronectin and hyaluronic acid (HA), and later collagen and proteoglycans. Fibronectin plays a major role in cell adhesion, growth, migration and differentiation. Fibroblasts secrete proteases including metalloproteases, which digest the plasma fibronectin, and then the fibroblasts secrete cellular fibronectin and assemble it into an insoluble matrix. Fibronectin links the components of the ECM to one another and to the cells.

7

Collagens synthesized by fibroblasts and provide strength to the tissue. Collagen synthesis in induced by PDGF, basic fibroblast growth factor, TGF- β, interleukin (IL)-1 and tumour necrosis factor. Proteoglycans, such as heparansulfate, keratansulfate, chondroitin sulfate, hyaluronic acid, etc. help to regulate the structure and permeability of the ECM and can modulate the growth and differentiation of cells. Re-epithelialization occurs upon the migration of cells from the periphery of the wound to the centre and is initiated within one day of injury; the division of peripheral cells occurs between 48-72 h, resulting in a thin epithelial cell layer which bridges the wound (Honnegowda *et al.* 2015). A moist environment is very important during this stage as it accelerates the process and the epithelial cells finally differentiate until they form a continuous layer of the epidermis (Junker *et al.* 2013).

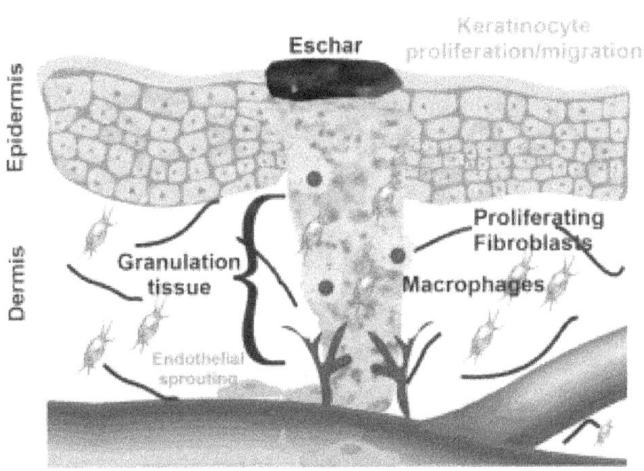

Figure 1.4 Proliferative Phase

1.1.1.4 Phase 4: Maturation or Remodeling Phase

This is the last and final phase that starts from week 3 and can last up to a year or more. It is the process of remodelling the collagen fibers laid down during the proliferation phase. Blood vessels formed in the granulation tissue are not required during this phase of healing and are removed by apoptosis. Type III collagen, a soft gelatinous collagen laid down in the proliferation phase, is replaced with the more highly structured type I collagen (Thiruvoth *et al.* 2015). The tensile strength continues to increase up to 80 % of normal tissue during this process. The differentiation of collagen is a dynamic process and although it takes place predominantly during the maturation phase it may continue to be remodeled indefinitely. The tissue is realigned along the lines of the stress and this process is mostly regulated by PDGF and TGF-ß and fibroblast growth factors. Maintaining a delicate balance between degradation and synthesis is required for normal wound healing which is controlled by regulatory mechanisms. Hyaluronic acid and fibronectin are degraded as collagen bundles become organized. Metalloproteinase activity decreases via the action of tissue inhibitors upon the accumulation of new tissue. Finally, a matured scar with a decreased number of cells and blood vessels, which has a high tensile strength, is formed. Since remodeling is a long process, closed wounds can quickly breakdown and reopen if the initial causes of the wound were not properly addressed (Paul & Sharma 2015).

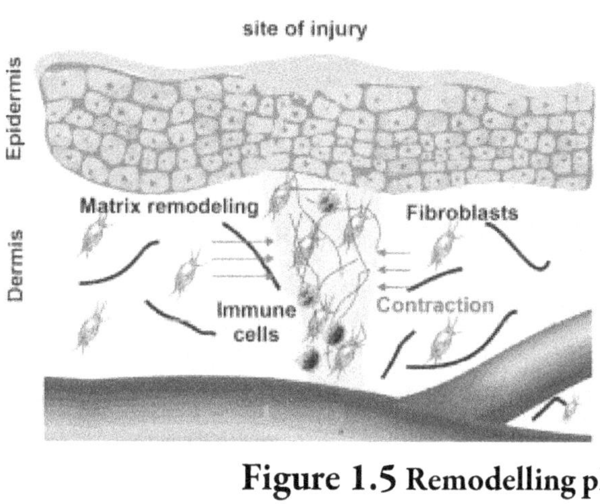

Figure 1.5 Remodelling phase

1.1.2 Comparison of Acute and Chronic Wound Healing

Acute wound healing is a sequential and dynamic process which consist of four explicit and overlapping phases such as hemostasis, inflammatory, proliferative and remodeling phase. In non-healing chronic wounds this process has been lost due to the absence of the hemostasis phase, causing ulcers that leads to the state of chronic inflammation that is self- sustaining and characterized by abundant neutrophil infiltration with associated reactive oxygen species and destructive enzymes (Thiruvoth *et al.* 2015).

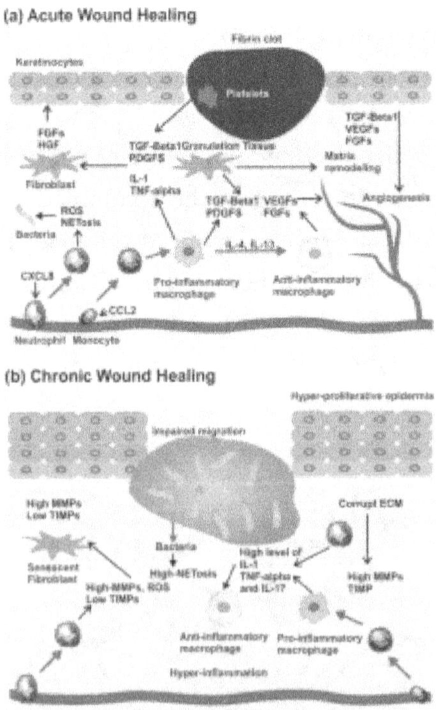

(Source: Arthe & Arivuoli 2020)

Figure 1.6 Acute Versus Chronic Wound Healing

Microenvironment in the normal wound bed is described by the presence of various growth factors, a systematic ECM and reactive cell populations. Matrix synthesis exceeds its degradation and MMP activity that is regulated by MMP inhibitors (TIMPs). Angiogenesis and neovascularization of wound proceeds in a timely behaviour via well-regulated sprouting of existing blood vessels and recruitment of endothelial progenitor cells (EPC) respectively. Finally, unlike their chronic counterparts, acute wounds are generally characterized by low bacterial burden.

Chronic wounds often have high incidence of bacterial film, leading to persistent inflammation, excessive proteolysis and degradation of growth factors, receptors or ECM. Cells residing within these wounds are unable to proliferate and migrate efficiently because of the absence of functional receptors or promigratory matrix substrates. Impaired angiogenesis and neovascularization both are the hallmarks of chronic wounds that results in insufficient oxygen and nutrient supply for the cells residing within the wound bed leading to further wound bed mutilation and impaired healing (Demidova *et al.* 2012). Healing will only occur after this prolonged inflammatory phase is broken and the wound micro environment is once again in balance thus allowing the wound to proceed to the next stages.

1.1.3 Complications in Chronic Wounds

Many physiological events and environmental agents can divert this healing cascade, thereby causing complications in the healing process. The complication in wound healing can be classified as delayed or non-healing, poor healing and uncontrolled healing. The Non-healing wound may be due to disruption of the normal physiological processes of repair that extends the inflammatory phase without the progress of the proliferation phase. The poor healing may be due to an improper inflammatory phase leading to the poor proliferation and inadequate collagenisation. Whereas in the uncontrolled healing is the collagenisation led to the excessive scarring, hypertrophic scars and keloids (Ehrichman *et al.* 1991; Paul 2015).

The non-healing wound is the wound that has failed to progress through the phases of healing in an orderly and timely fashion and show no significant healing for even a month. Chronic wounds affect the epidermis and dermis, or all the way to the fascia (Han & Ceilley 2017; Iqbal *et al.* 2017). Factors responsible for the chronicity of the wound are pressure, trauma, debilitating illness, increased bacterial load, Ischemia, excessive proteases, degraded growth factors, matrix metalloproteinases (MMPs), degraded cell surface structures, Senescent or aberrant cells, inappropriate treatment (Guo & DiPietro 2010). Ischemia is one an essential factor that causes inflammation in tissues and cells to release factors like interleukins, chemokines, leukotrienes and complement components that attract neutrophils (Sivler *et al.* 2018). Neutrophils remains longer in chronic wounds contributing to the lengthened inflammatory phase causing elevated levels of proteases such as matrix metalloproteases, plasmin, elastase and thrombin, that destroy components of the extracellular matrix (ECM) thereby damaging the growth factors and their receptors that are essential for progression of healing as shown in the Figure 1.6. (McCarty & Percival 2013).

Also the elevated levels of oxygen free radicals alter these essential components, that plays an important role in the pathogenesis (James *et al.* 2003). The matrix is an important component of the proliferative phase of an acute wound and also act as an essential medium for the cellular interaction in healing the injury which is "corrupted" in chronic wounds (Schultz & Wysocki 2009). It has been showed that the fibronectin is totally degraded in wound fluid collected from venous ulcers and is partially degraded in the exudates collected from chronic diabetic foot ulcer (Loots *et al.* 1998).

12

Interestingly, the addition of plasma fibronectin to the chronic wounds resulted in rapid degradation of fibronectin (Heerick *et al.* 1992).

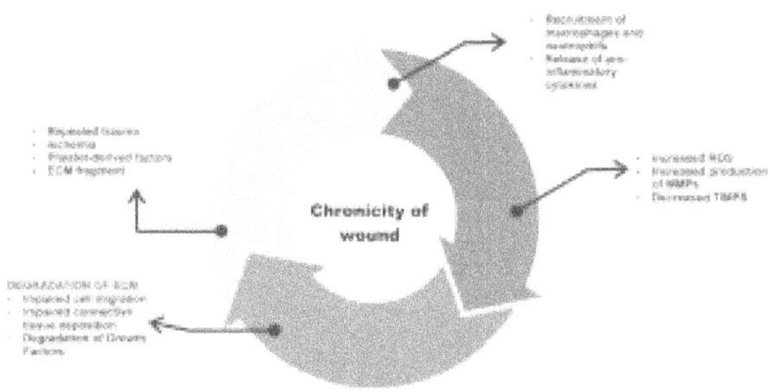

(Source: Arthe & Arivuoli 2020)

Figure 1.7 Inflammation cycle that contributes to the chronicity

The Extracellular matrix plays a vital regulatory role in the biological behaviour of epidermal cells. The epidermal cells must be in contact with early matrix containing fibronectin, fibrin and collagen to migrate from the wound edges (Olczyk *et al.* 2014; Tracy *et al.* 2016). These Migrating cells move from the edge toward the centre adhering to the ECM traveling along the fibrin, fibronectin and collagen strands (Lenselink 2015). It is justifiable that degradation of the protein by high proteolytic activity, contributes to the delayed proliferative phase (Schultz & Wysocki 2009). Typically, in chronic wounds the epithelial cells proliferate on the wound edges but fail to migrate. Lack of a normal ECM at the wound edges is now known to be the real problem. Failure of re-epithelialization is therefore an excellent example of a matrix deficiency state. Evidence of collagen degradation has been inferred from data measuring levels of MMPs in chronic ulcer fluid. Analysis of tissue biopsies from chronic ulcers showed that the levels of MMP are significantly higher and are associated with inflammatory components when compared with normal skin tissue. Other proteases also play an essential role in damaging ECM components and growth factors (Frykberg & Banks 2015). The serine protease urokinase plasminogen activator (uPA) activates a proteolytic cascade, resulting in the activation of MMPs. The lack of regulation of protease activity in chronic wound leads to the degradation of ECM thereby eventual tissue breakdown, ulceration and skin stripping seen in venous leg ulcers and pressure ulcers (Turk 2006).

The fibroblasts, keratinocytes and endothelial cells in a chronic wound go through several cycles of replication, probably reach their limits of proliferation and become senescent. Evidence of poor proliferative potential is obtained from the experimental observation of fibroblasts cultures from older animals, which exhibit decreases proliferative potential *in vitro* (Falanga 2003). Indirect evidence suggests decreased proliferative potential of fibroblasts from diabetic ulcer and venous ulcer fibroblast compared to normal fibroblasts (Bhushan *et al.* 2002) though the reasons are for this are unclear. Thus the characteristics of a typical chronic wound are an autonomous state of inflammation, a corrupt ECM, poorly responding cells particularly fibroblasts limiting ECM production, failure of re-epithelization due to lack of ECM orchestration and a defective scaffold (Rousselle *et al.* 2019).

1.1.4 Biomimetic Approach for Treating Chronic Wounds

The wound healing is an intricate process that results from a specific interaction between cells (platelets, leukocytes, fibroblasts, keratinocytes and endothelial cells), cytokines (TGF-β, PDGF, TNF-α, VEGF and FGF) and ECM. But in Chronic wounds, there will be low amount of connective tissue and ECM that needs an extra material to mimic the ECM to promote the wound healing process. The development of biomimetic material is a favourable approach for chronic wound healing into a regenerative process

(Chantre *et al.* 2019). Every tissue inside the body has a unique set of cells and ECM proteins arranged into a distinctive architecture, thus requiring

14

the properties of bioengineering materials to be designed in an organ-specific way (Bonnans *et al.* 2014). The lengthen inflammation and higher level of Matrix Metalloproteniases (MMPs) at the wound site causes significant degradation of ECM that delays wound healing process leading to chronicity. Thus the development of therapeutic dressing to control and positively regulate MMPs balance helps in achieving faster healing. The design of biomaterial matrices has the challenge to mimic the function of ECM that helps in fibroblast migration at wound site (Ramanathan *et al.* 2018). Recently, natural polymers have highly attracted the scientifical community interest. By knowing the biocompatibility and biodegradable nature of the naturally occurring proteins and polysaccharides helps in highest level of biomimicry, replicating the biological and physico-chemical features of the native ECM. By looking into our natural surroundings and by re-using some of the discarded natural resources, several functional biomaterials can be easily identified and implemented for promising wound healing applications, with a reduced impact on the environment. Nature itself can be better inspiration to develop economical, reduced energy consumption and fully biodegradable materials, providing great environmental sustainability. The increase interest in the use of either protein-based or polysaccharide-derived dressings provides striking and reflects the growing approach of giving back what we borrowed from Nature (Suarato *et al.* 2018). However, the natural polymer dressing has some limitations like weak mechanical strength and quicker degradation rate which can be improved through blending or mixing with other natural or synthetic polymers. Naturally polymers and their derivatives provide a versatile, multitasking and tunable platform to design appropriate extracellular environments that actively contrast the onset of infection and inflammations, while promoting tissue regeneration and scar remodeling.

1.1.4.1 Role of ECM in Wound Healing

The extracellular matrix is the largest component of normal skin that provides compressibility elasticity and tensile strength. The ECM consists of three classes of molecules such as structural proteins-collagens and elastins, protein-polysaccharide complexes to embed the structural proteins- proteoglycans and adhesive glycoprotein to attach cells to matrix – fibronectin and laminins (Hayden *et al.* 2005). In healthy skin, the extracellular matrix helps in supporting the cells and comprises key components of the basement membrane, which anchors and helps refreshing the epidermal cells.

Extracellular matrix constituents are essential components of the wound repair phenomenon. The ECM components create "scaffolding" that is essential in the repairing process that provides way to the structural stability of the matrix during each stage of the healing process (Raghow 1994). On the other hand, the ECM components are connected with the action aspects of the healing processes as these components fulfil a function of signal transduction and interactive sequence of biological reactions (Agren & Werthen 2007). The latter functions are connected with stimulation the adhesion and migration of cells during the healing process as well as with the mediating the interactions among cells, between cells and the matrix or between ECM proteins (Tran *et al.* 2004). ECM components also serve as a reservoir and a modulator of cytokines and growth factors thereby regulating wound repair activity. When the ECM is dysfunctional, wound healing slows or stalls. Chronic wounds contain increased level of inflammatory cells, giving rise to elevated levels of proteases that degrades the ECM components, growth factors and receptors that are essential for healing (Barrientos *et al.* 2008). Thus the importance of restoring the functional ECM in chronic wounds has led to technical advances and development of products that reduce excessive protease levels to contribute functional ECM proteins, thereby facilitating the healing process.

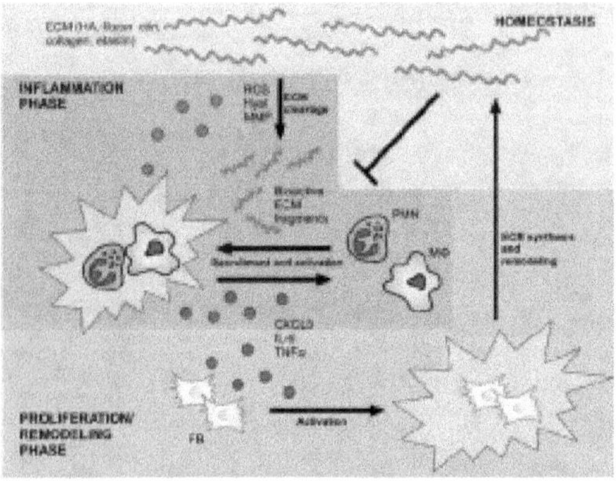

(Source: Salbach *et al.* 2012)

Figure 1.8 Mechanisms during normal and abnormal ECM turnover in the skin

1.1.5 Importance of Self Treatment in Chronic Wounds

Chronic wounds have a large impact on patients, including pain, loss of independence, long-term immobility and isolation, as well as financial costs associated with rehabilitation, nursing services and wound dressings (Hunt *et al.* 2000). The self-treatment of chronic wounds is a part of a self- care approach which mainly focuses on the ability of the individuals to promote and maintain the health condition by coping up with the illness and disabilities with the support of healthcare providers (Webber *et al.* 2013). The self-treatment of chronic wounds involves wound cleansing, wound inspection, applying and removing wound dressings by themselves. Although self-treatment of chronic wounds has not been well in practice, little research has been done on the self- care of chronic wounds (Kapp *et al.* 2017). However, the self-care treatment has been successfully improving symptom control, satisfaction and quality of life in other areas of chronic disease (Kapp & Santamaria 2017). Hence promoting self-care can empower patients, reduces the nursing workloads and cost burden mainly in the current pandemic situations. Thus, the present study aims in developing bioactive wound dressing that helps in self-treatment of chronic wounds.

1.1.6 Features of Ideal Wound Dressing

A dressing is chosen based on the depth, type, location and extent of the wound, the amount of discharge/infection and wound adhesion. Traditional wound dressings such as cotton bandage/gauze absorb a big part of the moisture containing of wound which dries the surface of wound intensely and leads to decreasing healing rate and pain in separating the dressing (Ghomi *et al.* 2019). In contrast, a broad range of polymers in the form of films, foams and gels have been developed that provides an optimum condition in the wound healing process. An ideal wound dressing should have the following 10 characteristics such as (1) moisture control (2) good transmission of gases, (3) elimination of excess exudate (4) protection against infection (5) decreased surface necrosis (6) mechanical support (7) easy removal (8) biocompatible, biodegradable and non-toxic (9) relieve the wound pain and (10) coft effective (Kokabi *et al.* 2007).

1.2 NOVELTY OF THE WORK

Chronic wounds especially diabetic foot ulcers are severe and common complication in patients with diabetes mellitus with the cumulative incidence of upto 25%. These ulcers cause substancial morbidity, impair quality of life, most importantly risk for lower-extremity amputation, higher treatment costs. Also, during this pandemic crisis, there were many diabetic patients struggled for getting treated in the hospitals during the first Covid-19 Lockdown. Thus, in this study, we aim to develop a self-care dressing which can be used as a bandage by patient themselves with the advice of the Doctors. The source of interest here is a beta-glucan polysaccharide specially the derived from *Euglena gracilis* called paramylon as its yield is more compared to that of other beta-glucan sources. The beta-glucans are known to have promising effect on metabolism and immune response that enhances the wound repair by increasing the infiltration of macrophages stimulating tissue granulation, collagen deposition and reepithelialization. Eventhough many researchers worked on the different sources of beta-glucan on wound healing, only few researches have been done using paramylon. Also, as far as we are aware of only limited studies has been done to develop a dressing material based on paramylon. There are studies to prove the wound healing activity of paramylon but to develop as product or dressing material there are only few researches developed. At the initial stage of the studies, we tried creating materials with different kinds of polymers with paramylon. But due to the solvent used and robust mechanical properties, silk fibroin has been used further with paramylon for the development of various dressing materials. As far as we know, there was no studies reported before with the combination of silk fibroin and paramylon so far even for any other application studies. Preliminarily, we were successful in preparing blend films of silk fibroin and paramylon with formic acid as the solvent. Indeed, at present, general research in the development of self-care dressing bandage is still in its infancy for chronic wounds as it requires wound cleansing, wound inspection, applying and removing wound dressings and/or applying and removing other therapies such as compression bandaging in the later stage. But this case can be neutralized if the chronic wounds are treated with the significant bandage during the early stage or at the time of the injury itself.

1.3 AIM AND OBJECTIVE

The aim of this study is to develop self-care biomimetic and instructive wound dressing for chronic wound healing. The objective of the study is as follows

- To study the biocompatibility of silk fibroin and Paramylon

- To develop various forms of dressing based on silk fibroin and paramylon

- To characterize all forms of dressing to optimize the better dressing for developing a self-care wound dressing bandage to treat chronic wounds

1.4 OUTLINE OF THE THESIS

The next **Chapter 2** reviews on the biomaterials particularly about the natural biomaterials and the components of extracellular matrix and importance of protein and polysaccharide complex in mimicking the role of extracellular matrix. This chapter contains the brief review about the source of silk fibroin and beta-glucan and the selection of silk fibroin from Bombyx mori and beta-glucan from *Euglena gracilis* for accelerating the chronic wound healing. As a preliminary study, this chapter also contains the biocompatibility studies of paramylon blended silk fibroin films in the 5 different variations. Various biological compatibility studies like blood compatibility, cell viability and *in vitro* enzyme degradations were studied for further processing of the blends to prepare various forms of dressing for developing self-care wound dressing.

The **Chapter 3** deals with an introduction of Nonwoven materials and bioactive wound care management along with silk fibroin and its importance and versatile functionalizing property. It also discusses about the fabrication methods of silk fibroin based nonwoven mat functionalized with paramylon of 0.25 %, 0.5 % and 1% for chronic wound healing. All the nonwoven mats were studied to their characteristics by various analysis like Scanning electron microscopy, contact angle measurement for knowing the wettability, porosity, swelling capacity and biological characteristics such as hemolytic analysis for knowing the blood compatibility, *in vitro* enzyme degradation capacity and *in vitro* cell viability studies by MTT (3-(4,5- Dimethylthiazol-2-yl)-2,5-Diphenyltetrazolium Bromide) assay for knowing its toxic nature. Along with this the functionalized mat that is having a better property in comparison with the other studies was used further for knowing the clotting nature using dynamic blood coagulation assay and *in vitro* wound scratch assay for knowing the wound healing ability. It was concluded that the all the silk fibroin based nonwoven mats was found to have a network of fibers with interconnected voids and were porous and white in colour. The wettability, swelling capacity and degradation properties showed that the paramylon functionalized silk fibroin fiber based non-woven mat showed better results than silk fibroin nonwoven mat. From the blood compatibility and cell viability testing, 0.25% paramylon functionalized nonwoven mat showed good biocompatibility and hemocompatibility which was found to have good wound healing activity with 92-99 % cell migration.

The **Chapter 4** deals with the chronic wounds that is having excessive exudates. This chapter strongly adhere on the absorption of chronic wound fluids that consists of metalloproteases that is responsible for the prolonged inflammatory phase thereby halting the wound healing process in chronic wounds. These exudates play a vital role in making the wound chronic and if uncontrolled, it even reduces the performance of the technology advanced bioengineering products. To regulate excessive exudate and to accelerate the wound healing process, wound dressing should possess high absorption and wound-healing properties. The cellulose sponge is economically efficient, which is highly porous with high absorbing and swelling properties. Silk fibroin (SF) is a natural non-mammalian protein, which is well known for its robust mechanical strength and biocompatibility that is functionalized with the paramylon to develop an instructive wound dressing material. Paramylon is a linear beta-1,3-glucan polysaccharide that has high bioactive functionality in the immune system as it can act as an immuno-modulator. Thus, in this chapter, Silk fibroin and paramylon blends were coated on cellulose sponges to develop a bioactive wound dressing with a high absorption capacity. For effective coating of blends on the cellulose sponge, a natural mordant called Potash alum is used as a fixative in the textile industry for natural fibers. The surface characteristics, porosity, swelling/water absorption capacity, blood and biocompatibility were evaluated. Along with this alum load silk fibroin/Paramylon coated on cellulose sponge was further used to know the clotting nature using dynamic blood coagulation assay and *in vitro* wound scratch assay for knowing the wound healing ability as it shows better biocompatibility compared to others.

The **Chapter 5** discusses the challenges in preparing the paramylon blended silk fibroin sponge for chronic wound healing. This chapter also fabricates the silk fibroin sponge sheet on which paramylon was coated to test its functionalization on the sponge sheet with the help of alum for fixation again. 0.5% paramylon with 50% formic acid solution was used for coating as the higher concentration of formic acid dissolves silk fibroin sponge itself. As discussed in the previous chapters, the surface characteristics, porosity, swelling/water absorption capacity, blood and biocompatibility were evaluated. Along with other studies, dynamic blood coagulation assay and *in vitro* wound scratch assay was studied for alum loaded paramylon coated silk fibroin sponge sheet to know its clotting behavior and wound healing ability.

The **Chapter 6** is the summary of the thesis which also provides the suggestions for future work.

CHAPTER 2

STUDY ON BIOMATERIAL SOURCE AND ITS BIOLOGICAL COMPATIBILITY STUDIES FOR DEVELOPING WOUND DRESSING MATERIAL

This chapter deals with the source of biomaterials for developing biomimetic bioactive material for chronic wound healing. A detailed description of materials with its extraction methods and experimental procedure adopted in investigation of biological compatibility is presented in this chapter.

2.1 BIOMATERIAL

A biomaterial is a non-drug material that suits for the inclusion in systems that replace the function of the bodily tissues or organs (Rakhorst & Ploeg 2008). In broad sense, natural or synthetic materials used to support or completely take over the function of the non-functional tissue in the human body are called Biomaterials (Hasirci & Hasirci 2018). Biomaterials due to their intended use in a very complex environment need to fulfil various requirement such as biocompatibility, comparable physical properties, appropriate stability, similar chemical properties, easy processing and sterilizable, appropriate bioactivity and cost effective (Reddy *et al.* 2021). The Biomaterials are derived from a number of sources like natural materials, synthetic polymers, metals, ceramics and composites. The choice of biomaterial is an important factor as that is intended through its use as a "scaffold" guide the regeneration process inside the body. These materials should be capable of being in contact with the bodily fluids and tissues for long time with minimal adverse reaction to the body (Ige *et al.* 2012). But due to its similarity to macromolecular substances in metabolic pathway and less toxicity, Natural biomaterials are considered to be more liable for the present study.

A wide range of natural polymers are available for the production of scaffolds because of their characteristics and better mimicry of natural extracellular matrix (ECM). Natural polymers can be tuned in terms of their biological activity, porosity, charge, mechanical strength and degradation rate by changing the polymerization conditions or polymer concentrations or by introducing functional groups (Atala *et al.* 2018). ECM is generally composed of well-organized networks of polysaccharide and proteins that play major role in different tissues (Stavolone & Lionetti 2017). In the last few years, the design of scaffolds based on blends of proteins and polysaccharides as surrogates of the native ECM has emerged as a promising strategy for several tissue engineering applications like skin, cornea, peripheral nerve, cartilage and bone (Rosellini *et al.* 2018). The source of interest here is a beta-glucan polysaccharide specially the derived from *Euglena gracilis* called paramylon as its yield is more compared to that of other beta-glucan sources. The beta- glucans are known to have promising effect on metabolism and immune response that enhances the wound repair by increasing the infiltration of macrophages stimulating tissue granulation, collagen deposition and reepithelialization. At the initial stage of the studies, we tried creating materials with different kinds of polymers with paramylon. But due to the solvent used and robust mechanical properties, silk fibroin has been used further with paramylon for the development of various dressing materials.

2.1.1 Silk protein

Silk protein is a two parallel monofilament spun by matured larvae of silkworm *Bombyx mori* which is composed of 65-75% fibroin, 20-30% sericin and 5% wax, pigments, sugars and other impurities (Cao *et al.* 2013). Silk plays a significant role in the economic life of man for 4000 years ago. Even today natural silk is the 'queen of textiles'. The silk produced from the silkworms and spiders are the most important forms of silks (Joseph & Raj 2012). India continues to be the Second largest producer of silk in the world. Sericulture is the cultivation of silkworms to produce silk. It is also a labour-intensive and agro-based cottage industry that provides occupation to millions of people in the rural and semi-urban areas in India (Hamid & Bhat 2021). Raw silk is of two kinds such as mulberry and non-mulberry. The distinction arises from the rearing of silk worms either upon mulberry leaves or on other plants. Mulberry silk is produced mainly in Karnataka, West Bengal, Jammu & Kashmir, Tamil Nadu and Andhra Pradesh. Non-mulberry silk comprising Tasar, Eri and Muga are produced in Assam, Bihar, Chattishgarh, Jharkhand, Madhya Pradesh, Orissa, Manipur and West Bengal. The highly characterized silks are from the domesticated silkworm cocoons, Bombyx mori (Babu 2019). Silks are fibrous proteins that are spun to fibers by insects. The silk from the cocoon of B. mori contains at least two major fibroin proteins; light and heavy chains of 25 kDa and 325 kDa, respectively (Joseph & Raj 2012). The core fibers are encased in a sericin coat, which is a family of glue-like proteins that holds two fibroin fibers together (Figure 2.1b) to protect the growing worm. Thc silk fibroin is the fibrous form of protein, whereas sericin is an amorphous glue-like protein (Mondal *et al.* 2007).

26

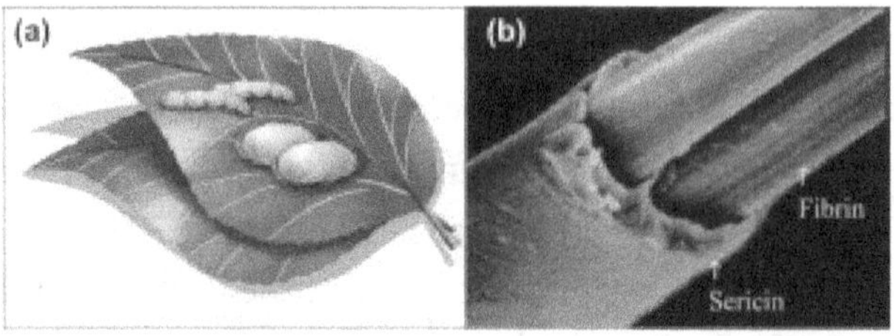

Figure 2.1 (a) Mulberry Silk Cocoon (b) Microscopic image of Silk Protein with the core Fibroins & Sericin coat

Silk sericin is a glue-like protein that holds fibroin fibers together. About 20-30% of the cocoon consists of sericin which are mainly composed of glycine, serine, aspartic acid and threonine (Das *et al.* 2021). For a long time, silk sericin was discarded in the silk industry through degumming process to improve its value as a textile fiber. In the field of regenerative medicine, due to its biodegradability, easy availability and hydrophilicity with many polar side groups, sericin is mostly copolymerized, crosslinked or blended with other polymers to form various scaffolds in order to improve the properties for relevant biomedical applications (Wang *et al.* 2014). Sericin was previously reported to elicit an immune response, which was later demonstrated to be triggered when sericin remains physically associated with fibroin (Jadhav *et al.* 2018). Considering the less mechanical property and its triggered immune response with silk fibroin, Sericin is removed completely from the silk cocoon in the present study.

Silk fibroin is a fibrous protein derived from Bombyx mori that has been highly used as a promising biomaterial in tissue engineering and regeneration. Compared with other biomaterials like polylactic acid and collagen, silk fibroin has exceptional mechanical strength, toughness and thermal stability (Holland *et al.* 2019). Silk fibroin is also containing the tripeptide Arg-Gly-Asp (RGD) sequences that can support cell adhesion, proliferation and migration of various cell types such as epithelial, endothelial, fibroblast, keratinocyte, glial and osteoblasts (Unger *et al.* 2004). The Silk fibroin was used as a suture for wound treatment centuries ago. Several studies reported that silk fibroin in different forms such as hydrogels (Ju *et al.* 2014), sponges (Roh *et al.* 2006), films (Padol *et al.* 2011), electro spun nanofiber mats (Schneider *et al.* 2009) and hydrocolloid dressings (Lee *et al.* 2016) have been used successfully as wound dressings to promote wound healing process.

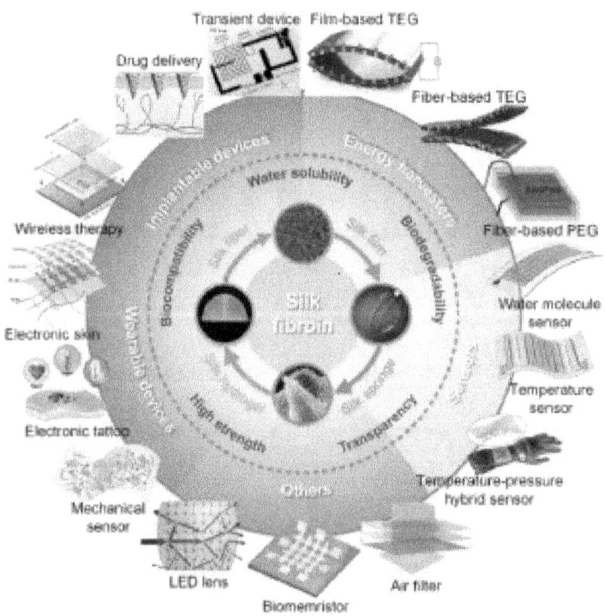

(Source: Wen *et al.* 2021)

Figure 2.2 Applications of Various forms of Silk Fibroin

2.1.2 Beta Glucan

"Beta glucan is the world most powerful immune booster known to man" (Hertzog 2013). Natural polysaccharides are abundant in nature, one of the most predominant class of polysaccharides is the β-glucans which are carbohydrate polymers that are found in the cell walls of many organisms such as bacteria, fungi, yeasts and some cereals like barley and oat (Ahmad *et al.* 2012). All β-glucans comprises of glucose polymer linked by 1-3 linear glycosidic chain core of varying length and branching structures (Aleem 2013). These branches that are derived from the glycosidic chain core are highly different with two main group of branching such as 1-4 or 1-6 glycosidic chains (Kaur *et al.* 2020). Also, different types of beta glucans exhibit distinct molecular weight, solubility and viscosity causing diverse physiological functions (Seo *et al.* 2019). It is also well-known for its powerful immune stimulant, antagonist of both benign and malignant tumours, anti-biotic properties and lower blood pressure or cholesterol levels (Bashir & Choi 2017). Since beta-glucan enhances the production of growth factor that are essential for skin, promotes collagen biosynthesis and maintains skins moisture and elasticity, we investigated the properties of various beta glucan in wound healing signalling pathway for boosting the wound healing process of chronic wounds (Majtan & Jesenak 2018).

The mechanism of beta-glucan is arbitrated by different receptors, especially Dectin-1 receptor (Brown & Gordon 2001), Toll-like receptor (TLR-2,4,6) (Underhill *et al.* 1999; Hsu *et al.* 2004), complement receptor 2 (CR3) (Ross *et al.* 1985), scavenger receptor (Rice *et al.* 2002) and lactosylceramide (Zimmerman *et al.* 1998). The Dectin-1 receptor is the most important as it is highly expressed in many immunocompetent cells such as dendritic cells (DC), neutrophils, eosinophils, macrophages, monocytes, several T lymphocytes and also in cutaneous cell (Willment *et al.* 2005). Upon binding with the Dectin-1 receptor, Beta-glucan stimulated many cytokines and activates immune and non-immune reactions (Sun & Zhao 2007). One of the potential applications of beta-glucan is dermatology as the topical application of β-glucan increases the anti-oxidant, anti-inflammatory, immunomodulatory effects, regenerative effects, moisturization and skin rejuvenation thereby helps in handling different skin disease and conditions via complementary therapy (Gautier *et al.* 2008; Jesenak *et al.* 2016). It also possesses anti-infective, anti-microbial activity against fungi, gram-positive and gram-negative bacteria (Jantova *et al.* 2015).

The Wound healing is an intricate process of different cellular and extracellular matrix cells and components that take part appropriately in an overlapping phase of wound healing (Landen *et al.* 2016). There are two modes of action possible in β-glucan activity of wound healing. One is by indirect influence of cytokines and macrophages and the other is direct proliferation of keratinocytes and fibroblasts. In Lengthen inflammatory phase of chronic wounds where macrophages in granulation tissue react with beta-glucan that acts as a source of growth factors and inflammatory cytokines such as IL-6, IL-1β and TNFα and the pro-inflammatory event is being mediated by the Dectin-1 receptor for normal aiding of wound healing. Numerous studies that are evident for the wound healing property of different type of beta-glucan has been reviewed in Table 2.1.

Table 2.1 ß-glucan types from various source and their respective wound healing activity

Name of the ß-glucan	Target cell type/ animal model	Inference
Baker's yeast Glucan	HDF	increased the nuclear factor-1 binding capacity and enhanced collagen biosynthesis (Wei *et al.* 2002)
	Porcine keratinocytes	enhanced the keratinocyte proliferation (Zulli *et al.* 1998)
	3T3 fibroblast	nanofibrous membranes enhanced the adhesion and proliferation of fibroblasts and keratinocytes (Wu *et al.* 2016)

	Venous ulcer biopsy	enhanced the healing in venous ulcer (Medeiros *et al.* 2012)
Barley Glucan	Adult Human dermal fibroblast (HDFa) / Mice	induces an early response in HDF cell favouring movement versus proliferation (Fuste *et al.* 2019)
Oat Glucan	Rats	increased antimicrobial activity, reduction of cholesterol and blood pressure (Shin *et al.* 2005)
Xyloglucan	Wister rats	exerted good healing effect in rats with severe wound (Hirose *et al.* 2019)
	NHEK, HaCaT and NHDF	promoted skin regeneration (Nie & Deters 2013)
Laminarin	Human corneal epithelial cells	enhanced the epithelial migration (Choi *et al.* 2013)
Paramylon	Mice	accompanied with a modest increase of inflammatory cytokines (Yasuda *et al.* 2018)
	HEK 293 T cells	acts as a bioactive supplement by boosting the cell proliferation capacity (Arthe *et al.* 2020)
Curdlan	Human Keratinocytes	stimulated the cell proliferation and migration in a Dectin-1 dependent manner (Van den Berg *et al.*

	2014)
Swiss 3T3 fibroblast & wister rats	Nanofibrous dressing of PVA/ curdlan incorporated with Ag has fast healing of wound in rats (Basha *et al.* 2018)

Table 2.1(Continued)

	BALB / cnude mouse	Membrane containing 50% β-glucan and Poly-(lactic co glycolic acid), accelerated the wound interactions (Kim *et al.* 2012)
	ddY mouse	Beta-glucan and chitosan complex enhanced the wound repair by activation macrophages and cytokine release (Kofuji *et al.* 2010)
β-(1,3–1,6)-D-Glucan from Aureobasidium pullulans	Human dermal fibroblasts	enhanced the dermal fibroblast migration and proliferation that modulated the effect of transforming growth factors (Choi *et al.* 2016)
	Human dermal fibroblasts, adipose tissue derived stem cells	boosted up the cellular response, migration and proliferation of both the cells (Woo *et al.* 2010)
Schizophyllan (SPG)	L292 Fibroblast	SPG based nanofibrous scaffolds showed cell proliferation and cell migration (Safaee-Ardakani *et al.*

		2019a)
		PVA/SPG-AgNPs nanofibers showed anti-microbial activity thereby helping in reducing the infection in the wound (Safaee-Ardakani *et al.* 2019b)
Lichenan	NHEK and HaCaT keratinocytes	stimulated human keratinocytes by specific mechanism into the terminal differentiation (Esch *et al.* 2019)

β-glucan currently used in nutritional supplements are from Saccharomyces cerevisae, but it is wide present in nature. It is also found higher in plants, bacteria, fungi and algae. (1,3)-β-Glucan and its related glucans are essential both as storage and structural polysaccharide in protozoans and chromistans (Delatte *et al.* 2001). Over the past decades, efforts have been taken towards the characterization of properties and functions of beta-glucan synthases from these organisms. But most of the beta-glucan synthases are membrane associated complexes that are difficult to purify and characterize due to their inherent instability after solubilization from their native membrane environment (Bacic *et al.* 2009).

2.1.2.1. Paramylon

β-glucans present in many organisms have the capability to activate the innate immune system. But, all the enzyme complexes producing beta- glucan in plants, bacteria and yeast are known to be present in the plasma membrane and secrete the beta-glucan either into the cell wall or outside the cell. Therefore, utilization by the biotechnology industry is difficult as such compounds are only often active when present within the membrane rather than in soluble fraction. The protist *Euglena gracilis*, however manufactures an intracellular granular reserve beta-glucan in high amounts (Drake *et al.* 1992). Also, Paramylon is one of the purest forms of crystallized β-(1,3)-glucan present in nature which is extracted from *Euglena gracilis*. Because this reserve carbohydrate is granular and the enzymes synthesizing would be soluble rather than present in the membrane and thus these glucan synthases would be easier to use from a biotechnological point of view. They could be expressed in microorganisms and the carbohydrate isolated after growth in bioreactors. According to the yield and extraction procedures, beta-glucan from *Euglena gracilis* is economical for the wound dressing application.

Paramylon (BG) is β-(1,3)-D-glucan extracted from *Euglena gracilis* which yields paramylon of about 60-70% of the total dried cell (Barsanti *et al.* 2001). Paramylon shows cytokine-related immunopotentiation (Kondo *et al.* 1992). Sulfated derivatives of paramylon and N, N-dimethyl aminoethyl paramylon exhibit anti-human immunodeficiency virus (HIV) and antimicrobial effects (Koizumi *et al.* 1993). Paramylon has been developed as a polysaccharide nanofiber, thermoplastic and as a transparent thin film and also similar to the bacterial cellulose, paramylon film can maintain a moist environment at the wound surface. To our best knowledge, only a very few literature is currently available on paramylon film efficiency on wound healing using an animal model (Shibakami et al. 2013 ; Yasuda et al. 2018). Yasuda et al. (2018) also concluded that the paramylon film regulates the excessive immune responses resulting in acceleration of wound healing.

2.2 DETERMINATION OF WOUND HEALING CAPACITY OF SILK FIBROIN AND PARAMYLON BLEND FILMS

The early use of temporary scaffolds in medicine was only for the simple mechanical support, whereas research efforts in recent decades have expanded the potential functional role that the scaffold can have a bioactive agent that helps in maintaining the structural or functional integrity of the host tissue (Atala *et al.* 2018). Due to the vast variety of natural or synthetic polymers, it is important to plan and determine the suitability of the blends to evaluate that it fulfils the key requirements such as biodegradability, biocompatibility and bioactivity (Barbeck 2019). For this purpose, we previously studied the physical and chemical properties of the pure and blend films preliminarily and observed structural, thermal, mechanical and other physical properties which resulted that the silk fibroin film has better thermal stability and stiffness property whereas the paramylon film showed good water absorption property (Arthe et al. 2020). Thus, to evaluate the key requirement, Silk fibroin (SF) and Paramylon (BG) Blends were prepared and studied further to know its degradation capacity, biocompatibility and wound healing activity.

2.2.1 Experimental Section

2.2.1.1 Materials

Bombyx mori silk cocoons were supplied by the Department of studies in Sericulture sciences, University of Mysore. Paramylon used was purchased from Sigma-Aldrich chemical Co.(St. Louis, MO.) and used without further treatment. All the other chemicals used were analytical grade obtained commercially. Human Embryonic Kidney (HEK) 293 cell line and L929 cell line (murine fibroblast) was purchased from National Centre for Cell Science (NCCS), Pune.

2.2.1.2 Extraction of Silk Fibroin from Bombyx mori Cocoon (bivoltine, white shell)

The *Bombyx mori* cocoons were cut into small snippets and boiled with 0.02 M Sodium Carbonate for 30 minutes under constant stirring for degumming to remove sericin from fibroin. The degummed fibers were washed up to 3 times with ultrapure water to completely remove the glue like sericin protein. The washed fibers were dried overnight in a sterile condition. The dried degummed fibers were dissolved in 9.3 M Lithium Bromide (LiBr) for 4 hours at 60°C, which was then dialysed against sterilized water for three days, changing water every 6 hours in order to remove LiBr. After dialysis, the regenerated silk fibroin solution was centrifuged to remove the debris and was kept under -20°C followed by lyophilization at -110 °C for 36 hours to form silk fibroin powder (Rockwood *et al.* 2011)

2.2.1.3 Preparation of Silk Fibroin and Paramylon Blend Films

The SF/BG blends were formed by making 1% solution by weight with 98% formic acid as a solvent. The solutions were prepared at five different ratios as 1/0, 0.75/0.25, 0.5/0.5, and 0.25/0.75, 0/1 of Silk fibroin to paramylon solution by weight forming five groups of test samples which are coded as SF, FG 75:25, FG 50:50, FG 25:75 and BG respectively. All the solutions were stirred continuously for 4 hours to form a clear solution before casting in polystyrene plates. The plates were then left in oven at 40°C for 48 hours to obtain a transparent film of SF, BG and FG.

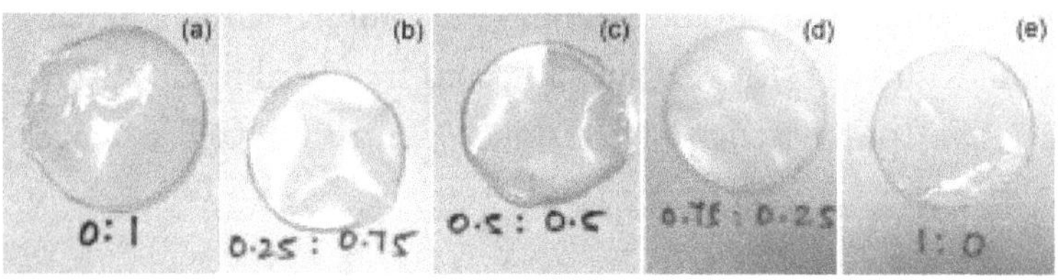

Figure 2.3 Appearance of SF, BG and Blend Films

2.2.1.4 Biological Compatibility and Degradation Studies

A. *In vitro* Enzyme Degradation Studies

Degradation of SF, BG & FG films were studied as percentage of weight remained after incubation in lysozyme enzyme solution for about 4 weeks. The films with known dry weight before degradation (W_0) were submerged in Phosphate Buffer Solution (PBS) of pH 7.4 containing 10μg/mL lysozyme and PBS alone as blank at room temperature for four weeks which was refreshed weekly. At various time intervals (1, 2, 3 & 4 weeks, the films were removed and dried overnight. The weight of the dried films is the weight after degradation (W_t). The percentage of the weight

36

remained after degradation was analysed using the following formula (Nwe *et al.* 2009),

% Weight remained = $(W_t / W_0) \times 100$ (2.1)

B. Blood Compatibility Testing - *In vitro* haemolytic activity assay

Haemolysis activity assay was performed to evaluate the blood compatibility of the SF, BG and FG films. In this experiment, a dry film was equilibrated in saline water (0.9% NaCl solution) for 24 hrs at 37°C. Healthy Human ACD blood (0.25ml) was added onto the wet films. After 20 min, 2.0 ml of saline was added onto the films to stop hemolysis and the test samples (TS) were incubated for 60 min at room temperature. Positive control (PC) and negative control (NC) were obtained by adding 0.25 ml Human ACD into distilled water and saline solution respectively. Incubated films were centrifuged at 4°C (1500 rpm, 5 min), the supernatant was taken and its absorbance at 545 nm was recorded. The percentage of hemolysis was calculated using the following formula (Kurhade *et al.* 2013)

Hemolysis (%) = $[(A_{TS} - A_{NC}) / (A_{PC} - A_{NC}] \times 100$ (2.2)

C. *In vitro* Cell Viability Studies for Cytotoxicity Analysis

Cell viability was determined using MTT assay. HEK 293T cells used in this assay were grown in regular high glucose Dulbeccos Modified Eagles Medium (DMEM) supplemented with 10% fetal bovine serum (FBS) and antibiotic-antimycotic mix. For this assay, the cells were resuspended in culture medium and seeded in control wells or wells containing test samples placed in a 96-well plate. The cells were then placed in a humidified atmosphere of 5% CO_2 for 24 hours or 72 hours. Following incubation, the medium was removed and each well was treated with 0.5mg/ml of 3-(4,5-

37

dimethylthiazol-2-yl)-2,5-diphenyl tetrazolium bromide (MTT) dissolved in DMEM, and incubated for 2 hours at 37 °C in a humidified atmosphere of 5% CO_2. The yellow MTT is reduced to blue-purple formazan by the mitochondrial dehydrogenase from intact living cells. The resulting blue precipitate was solubilised in Dimethylsulfoxide (DMSO) and the absorbance was recorded at 570 nm using a standard plate reader (Wiegand *et al.* 2019). Experiments were performed in a triplicate and the data were presented as mean±SD (n=3). The percentage of cell viability (CV%) was calculated using the formula,

$$CV(\%) = (Absorbance_{TS} / Absorbance_{control}) \times 100 \quad (2.3)$$

2.2.1.5 Wound healing Studies

To confirm the wound healing property of the silk fibroin and paramylon blends, one blend with significant property has been considered to for further analysis. Based on the cell viability studies, both FG 75:25 and FG 75:25 shows good proliferation rate, but our previous research studies showed that FG 75:25 showed good stiffness and thermal stability. Thus, from the above and previous studies, FG 75:25 have been considered further for analysing its blood coagulation property and wound healing activity along with the percentage of release of paramylon.

A. Dynamic Blood Coagulation Assay

Dynamic blood coagulation assay measures the release of haemoglobin from residual erythrocytes that remained free from entrapment during clot formation. The Assays were performed by recalcification of the anticoagulated whole blood. Briefly, FG 75:25 film was placed at the bottom of 100 ml beaker and then prewarmed in a water bath at 37 °C for 5 min. Eventually, anti-coagulated whole blood (0.25 ml) was dripped onto the surface of the film and incubated at 37 °C for another 5 min, after which

Calcium Chloride solution (0.02 ml of 0.2 mol/L) was dripped into the blood to initiate the coagulation cascade (time 0). The beakers were shaken for 1 min for uniform distribution of $CaCl_2$. Heating of the covered beaker was continued at 37 °C for a predetermined time (10, 20 and 30 min). At the designated termination point, the beaker was removed from the water bath and shaken for 10 min followed by addition of 50 ml distilled water to lyse the free erythrocytes. The absorbance of the supernatants at 540 nm was determined. The absorbance signal derived from the proportion of clot-free erythrocytes are inversely proportional to the size of the clots. The absorbance-time curve was constructed (Liu *et al.* 2014).

B. *In vitro* Paramylon Release Kinetic Assay

This Assay was performed to investigate the release kinetics (burst release vs. sustained release) of paramylon blended with silk fibroin. In order to evaluate the drug release kinetics, FG 75:25 films were cut into samples of average weight of 5 mg and then immersed in Phosphate-buffered saline for a period of 48 hours. At predetermined time intervals (5 min, 10 min, 30 min, 1,

2, 4, 8, 12, 24 and 48 hours), 3 ml of PBS was removed from the sample for detecting the paramylon release and the same 3 ml is replaced with fresh PBS at the scheduled times. UV Spectroscopy was used for detecting the paramylon release profile at 208 nm. At the beginning of the release test, a linear calibration curve of standard paramylon solution from 5 to 20 µg/ml was created. Paramylon release behaviour in the FG 75:25 blend films were analysed according to first order model (Cam *et al.* 2020).

C. *In vitro* Wound Scratch Assay

L929 cells were grown in 24 well plates at a density of 1 × 106 cells/ml and cultured until ~ 80% confluency. A small linear scratch was created in the confluent monolayer by gently scraping with sterile cell scrapper as per the method described by Liang *et al.* (2007). Cells were thoroughly rinsed with 1XPBS to remove cellular debris and treated with different concentrations of FG 75:25 sample. Cell proliferation was monitored at different time points: 0 h, 4 h, 18 h, 24 h and images of the migrated cells were taken at all different time points using digital camera (Nikon, Tokyo, Japan) connected to the inverted phase contrast microscope (Radical instruments, India).

2.2.2 RESULT AND DISCUSSIONS

2.2.2.1 *In vitro* Enzyme Degradation Studies

The degradation of SF/BG blend films was studied *in vitro* by lysozyme hydrolysis (Siritientong *et al.* 2013), and the remaining weight percentage was determined (Figure 2.4) with lysozyme enzyme and PBS as blank. The initial weight loss of about 10% in week 1 may be due to the partial hydrolysis of the film along with the absorption and syrinkage upon drying. But, there was a gradual decrease in weight upto 20% after the 2nd week from when the actual enzymatic weight loss happens leading to the disruption of the films at week 4.

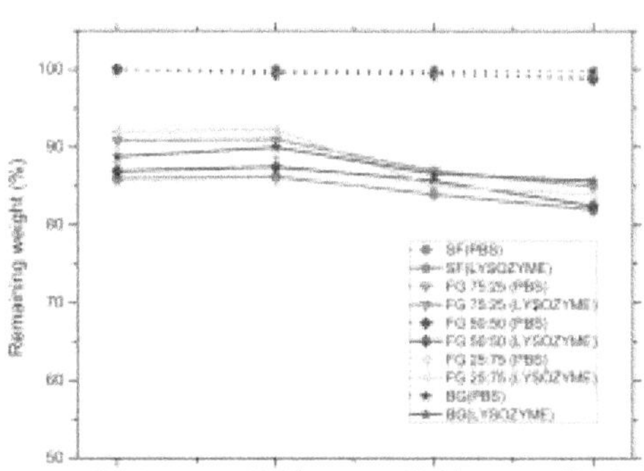

Figure 2.4 Weight remaining percentage of silk fibroin, paramylon (BG) and their blend films in enzymatic and non-

enzymatic degradation environment. Data represent mean±standard error of the mean(n=3)

2.2.2.2 Blood Compatibility Testing – *In vitro* Haemolytic Activity Assay

The Haemolysis activity of films was used to evaluate the critical interactions of films with blood to explore possible adverse effects arising from the exposure of the SF and BG to blood cells and proteins. Haemolysis rates of all the pure and the of blend films (Figure 2.5) were lower than 1.5% when compared with both positive and negative control thus confirming that the films show no negative impact on blood and have good blood compatibility as it does not show any traces of lysis in the blood cells (Panico *et al.* 2018).

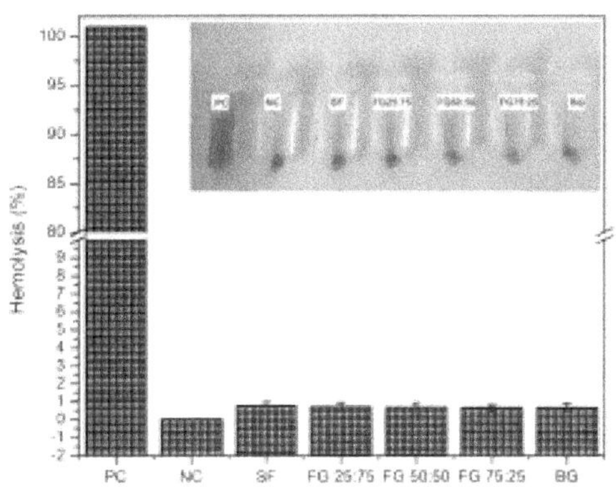

Figure 2.5 Percentage of Haemolytic activity in SF, paramylon (BG) and their blend films. Data represent mean±standard error of the mean(n=3)

2.2.2.3 Cytotoxicity Testing – *In vitro* Cell Viability Assay

To measure the toxicity in the films, we investigated the effect of pure and blended films on the viability of HEK 293T cells after incubation of about 24 and 72 hours by MTT reduction assay. As shown in Figure 2.6, the pure, as well as the blended films, did not show any noticeable toxicity as the cell viability remained above 90% at 24 hours and 95% at 72 hours of

41

incubation when compared to that of control at 24 and 72 hours respectively (Muniandy *et al.* 2018).

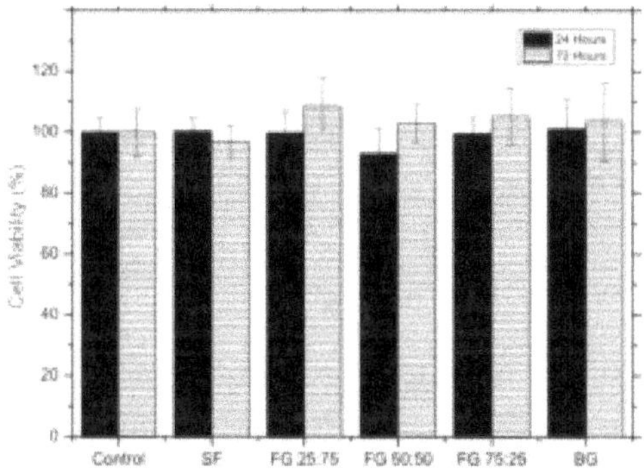

Figure 2.6 Effect of SF, BG and their blend films on HEK 293T cell viability. MTT assay of HEK 293T cells for 24 and 72 hours. Data represent mean±standard error of the mean(n=3)

2.2.2.4 Dynamic Blood Clotting Assay

The dynamic blood clotting time is an *in vitro* test that measures the degree of activation of intrinsic coagulation factors when surfaces interact with blood. Activation is measured by the absorbance-time curve for release of hemoglobin from erythrocytes that avoid being entrapped during clot formation (Liu *et al.* 2014). The curve (Figure 2.7) implies that the film shows similar pattern as on control and hence it does not affect the coagulation pathways. Despite this, the clot sizes appear to be bit higher in FG 75:25 films, since the magnitude of absorbance is inversely related to clot size.

42

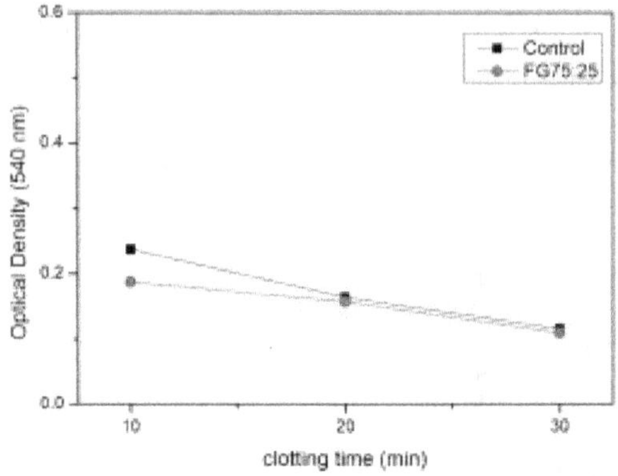

Figure 2.7 Intrinsic coagulation factors activation measurement curve

2.2.2.5 *In vitro* Paramylon release Kinetic Assay

In vitro drug release kinetic assays for FG 75:25 films were performed according to the first-order model. Primarily, the UV spectra were obtained with the concentration range of paramylon from 5 to 20 μg/ml and a linear standard calibration curve was drawn from paramylon absorption values (R2 = 0.99127) obtained for the quantitative determination from UV absorbance measured at 208 nm. The *in vitro* release studies were performed over a period of 48 hours for FG 75:25 films in order to determine the release kinetics of blended bioactive paramylon. The Films showed a burst release in approximately 60 minutes and maintained a sustained release after 8 hours (Figure 2.8). These results are consistent with previous studies in which the burst release reaches high concentration of bioactive molecule and a sustained release afterwards may help in wound closure (Cam *et al.* 2020).

43

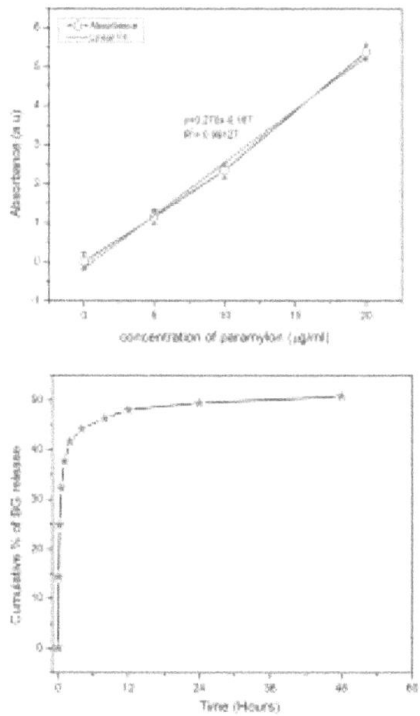

Figure 2.8 *In vitro* drug release profile of FG 75:25 film: (a) paramylon calibration curve (b) Paramylon release profile of FG 75:25 films according to first-order model

2.2.2.6 *In vitro* Wound Scratch Assay

Cell migration is an essential process in many physiological and pathological events such as wound healing, embryonic development and cancer metastasis. The basic procedure to study cell migration involves (i) creation of a cell monolayer, (ii) the production of a pseudo-wound in the confluent layer of cells, (iii) the capture of images at different time intervals until wound closure is reached, and (iv) the analysis of the image sequence in order to quantify the migration of the chosen cells (Cappiello *et al.* 2018). The *in vitro* scratch assay is an easy and economic method to mimic the cell migration during wound healing *in vivo*. L929 cells were incubated in presence and absence of the FG 75:25 films and images were captured at 0, 4, 18 and 24 h at various concentrations of FG 75:25 films (25, 50, 75 and 100 μg). From the Figure 2.9, it is clearly observed that the cell migration has started at the 4 hours after incubation in almost all the concentration and has reached upto 91-98 % within 24 h. Thus, confirming the wound healing capacity of FG 75:25 films on L929 Fibroblast cells.

44

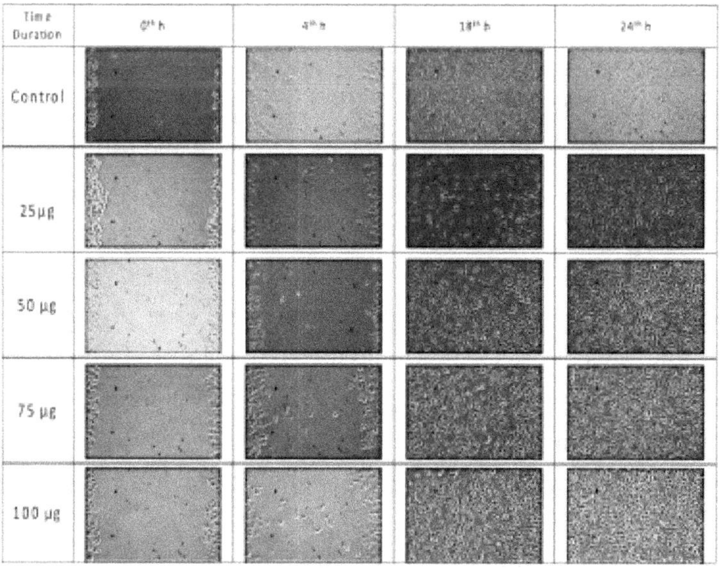

Figure 2.9 Scratch wound assay using L929 fibroblasts incubated with FG75:25 films (25, 50, 75 & µg) at 0, 4, 18, 24 hours

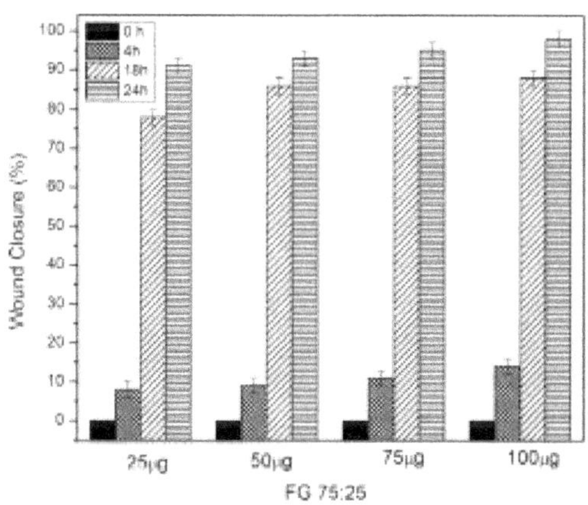

Figure 2.10 Percentage of wound closure at 0, 4, 18, 24 hours

2.2.3 Conclusion

This study provided a brief understanding of various properties of Silk fibroin and paramylon blends with respect to wound healing property. The SF, BG film and its blends were prepared using formic acid as solvent and various features such as biodegradation, biocompatibility and wound healing capacity were reported. From various physico-chemical studies previous reported and the biological studies, one blend FG 75:25 showed better property for developing bioactive self-care wound dressing. Hence, FG 75:25 was further studied to confirm the blood coagulation property and wound healing activity based on the release of paramylon using drug release kinetics. These FG 75:25 sample showed good migration rate at various concentrations such as 25, 50, 75, 100 μg. Thus, it was concluded that the FG75:25 film showed significant property for developing a bioactive self-care wound dressing and hence this can be further optimised for development of bandages that can be used by the patients by themselves and also this can be further studied *in vivo* with streptozotocin induced rats/mice.

46

CHAPTER 3

PREPARATION AND CHARACTERIZATION OF SILK FIBROIN PROTEIN BASED NONWOVEN MAT BLENDED WITH PARAMYLON FOR IMPAIRED WOUND DRESSING

This chapter reports on the fabrication of silk fibroin protein based nonwoven mat functionalized with the bioactive paramylon for impaired wound healing. A detailed procedures about the extraction of silk fibroin and fabrication of nonwoven mat was presented along with the various physical and biological characterization related to develop an ideal wound dressing.

3.1 INTRODUCTION

Nonwovens are engineered fabrics that are made directly from fibers or polymer granules that are bonded together in the form of mats. Nonwoven fabric is a technical textile that ranks third in the manufacture of textile surface after woven and knitted fabrics. The Nonwoven fabric has a number of advantages over woven and knitted as it can be designed with specific targeted characteristics, can be produced with substantial variations in thickness, mass, elasticity and stiffness and is comparatively fast and economical in manufacturing (Jones & Stylios 2013). There is broad variety of nonwoven processes. The processes basically consist of extruded fibers that are deposited or laid on a conveying surface in a dry, quenched in air, wet or molten state. The web formation phase of nonwoven processing transforms the fibers, together with or thermoplastic resins or films, into layers of loosely arranged networks such as webs, mats or sheets (McLoughlin & Sabir 2017).

47

Nonwoven fabrics are highly used in the medical field and in protection against biological agents. Shorter production cycles, higher flexibility and versatility and lower production costs makes the nonwoven materials more popular in medical applications (Bartels 2011). Most nonwoven products used outside the body are disposable and hence it is increasingly as a major component in the design of smart wound care products (Srikrishnan & Raghunathan 2013). These fabrics provide moist wound environment, controls vapour transmission and absorbency with low skin adhesion. In some cases, this fabric is subjected to needling to strength the fabric. The entanglement of fibers by needling will increase both wet and dry fabric strength and therefore the likelihood of the dressing will remain in one piece during the use and removal (Kellie 2016). Thus, different nonwoven structures of natural or synthetic origins were used in various medical applications like wound healing or cartilage, bone, nerve and ligament repair and regeneration. Also, increasing levels of skin infections and injuries such as pressure ulcers and diabetic foot are the primary factors that contribute to the development of the new trend called bioactive wound management (Okur *et al.* 2020). Bioactive composite wound dressing is an active area of research in the field of medicine with plenty of new and developing materials & methods. This bioactive wound dressing is produced from both artificial and natural materials that have fewer chance of infection, faster healing process, low risk of leaking that reduces the pain and discomfort and also cost and time effective. The Bioactive dressings are the one which delivers bioactive substances for the wound healing process either by delivering the bioactive compounds directly or activate them via endogenous activity (Rajendran 2009). This dressing belongs normally to the FDA dressing category of interactive wound and burn dressings (Georgescu *et al.* 2017).

3.1.1 Silk fibroin fibers

Natural polymers, such as proteins, polysaccharides or polynucleotides have better interactions with cells owing to their bioactive properties (Dhandayuthapani *et al.* 2011). Hence, Silk fibroin - a natural protein that is derived from the cocoons of domestic silkworm, *Bombyx mori* that is free from microbial contamination was used. Silk threads have been used from ancient days as a suture in post ocular, neural and cardiovascular surgery (Altman *et al.* 2003). Due to its robust biocompatibility, non-toxicity, non-carcinogenic activity and biodegradable properties (Sultan *et al.* 2018), Silk fibroin (SF) dressing has been applied in different forms such as hydrogel (Guan *et al.* 2020), film (Aparna & Mamatha 2021), electrospun mat (Ju *et al.*2016) and sponge (Roh *et al.* 2006) that could support the proliferation and adhesion of fibroblasts and keratinocytes. Also the versatile functionalization nature of SF helps in using it for innovative applications (Yan *et al.* 2012). Thus, Silk fibroin is functionalized with a bioactive molecule called β-glucan for treating chronic wounds.

3.1.2 β-glucans

β-glucans are glucose polymers derived from variety of sources including yeast, grains, fungus and algae. It belongs to the class of drug known as biological response modifiers due to its ability to enhance the immune system (Bohn & BeMiller 1995). On comparing with other sources, β-(1,3)-glucan from Euglena called Paramylon has a high molecular weight and when cultivated under optimal conditions, yields upto 90% of the cell mass (Russo *et al.* 2017). Paramylon also possess anti-atopic dermatitis effect (Sugiyama *et al.* 2010), immuno-potentiating activity (Nakashima *et al.* 2017), and hepato-protective effects (Sugiyama *et al.* 2009) and antimicrobial activity (Sakagami *et al.* 1991). Due to its high yield, inexpensive extraction methods and immune-potentiating activity, paramylon derived from *Euglena gracilis* is functionalized with silk fibroin fiber for developing a nonwoven mat dressing as it can be produced directly from fibers, disposable, light weight, versatile in the sense of bulk and area densities, flexible, and low production cost.

3.2 EXPERIMENTAL SECTION

3.2.1 Materials

Bombyx mori silk reeled filament was kindly supplied by the Avinaash Silk, Salem. β(1–3)-D-glucan used were derived from *Euglena gracilis* and was purchased from Sigma-Aldrich Chemical Co.(St. Louis, MO.) and used without further treatment. All other chemicals used were analytical grade obtained commercially. L929 cell line (murine fibroblast) was purchased from National Centre for Cell Science (NCCS), Pune.

3.2.2 Methodology

3.2.2.1 Preparation of degummed silk fibroin fibers

The reeled filament of *Bombyx mori* silk was treated with a boiling solution of 0.02 M Sodium Carbonate for 30 min under constant stirring for removing the core protein sericin from fibroin. Then the degummed fibers were washed two to three times with ultrapure water to remove the glue-like sericin protein and dried overnight under sterile conditions to get the degummed silk fibroin fibers (Rockwood *et al.* 2011). The fibers before and after degumming was examined using Carl Zeiss MA15/EVO 18 SEM to confirm the complete removal of sericin after degumming.

3.2.2.2 Production of paramylon blended Silk fibroin-based Formic acid-crosslinked 3D Nonwoven mat

Briefly, the degummed silk fibers were cut into small snippets and soaked at room temperature in 98% formic acid (FA) containing 0.01% w/v $CaCl_2$ in the material to fluid ratio of 1:200. This fiber suspension was shaken for approximately 30 min to form a homogenous fiber distribution and was evaporated under atmospheric conditions. The resulting silk fibroin non- woven mat (SM) was washed repeatedly with double distilled water to remove the residual salts and dried under vacuum (Pra *et al.* 2006). For functionalization, paramylon was added to the FA containing $CaCl_2$ suspension to form 0.25, 0.5, and 1% of solution in which small snippets of degummed silk fibers were added and the above steps of SM mat were repeated to form paramylon functionalized silk fibroin based non-woven mat. The paramylon functionalized mat was named as SBM25, SBM5, SBM1 based on the percentage of paramylon solution prepared such as 0.25%, 0.5% and 1% respectively. $CaCl_2$ used here as a Chaotropic agents that helps in partial dissolution of degummed fibers to form nonwoven mats.

3.2.3 Characterization Studies

3.2.3.1 Morphological studies by scanning electron microscopy

The Surface morphologies of the nonwoven mat was examined using Carl Zeiss MA15/EVO 18 SEM coupled with energy-dispersive X-ray (Oxford Instruments NanoAnalysis INCA Energy 250 Microanalysis System (EDX). About 0.5 cm of the dry sponges were cut and sputter coated with gold to enhance the surface conductivity.

3.2.3.2 Fourier transform infrared (FTIR) spectroscopy analysis

Secondary structure and conformational changes were studied by FTIR spectroscopy with air as the reference and in the spectral range of 4000

- 400 cm-1 at 4 cm-1 spectral resolution with 32 scans using JASCO International Co./Japan Fourier transform infrared Spectrometer Model FTIR- 6300 instrument.

3.2.3.3 Surface wettability testing - contact angle measurement

The static contact angles of all the nonwoven mats were measured at room temperature (RT) by sessile drop method using an optical contact angle meter to examine the surface wettability. The glass microscope slide was used as the substrate. Each slide was cleaned thoroughly and were soaked in ethanol overnight. 7 μL of distilled water was carefully injected on the nonwoven mat surface before measuring with the measurement time of 24 seconds and then the contact angles were measured (Mates *et al.* 2014).

3.2.3.4 Porosity measurement by solvent replacement method

Dried nonwoven mats were submerged in absolute ethanol overnight and weighed after blotting the excess ethanol on the surface. The porosity was calculated as per the following formula (Shabir *et al.* 2017),

Porosity = $[(M_2 - M_1)/\rho V] \times 100$ (3.1)

where, M1 and M2 are the weight of the mat before and after immersion in absolute ethanol respectively, ρ is the density of absolute ethanol and V is the volume of the nonwoven mat.

3.2.3.5 Swelling capacity testing

All dried nonwoven mats of known weight (W_d) were immersed in both PBS and water at room temperature for at least 24 hours. After swelling, the swollen weight (W_s) was measured after dabbing the excess surface water onto the tissue paper. Then, the ratio of swelling was calculated with the following equation (Larraneta *et al.* 2018),

$SI = [(W_s - W_d) / W_s] \times 100$ (3.2)

3.2.3.6 *In vitro* enzyme degradation studies

The biodegradation rate of all the nonwoven mats was determined *in vitro* by measuring the change in the weight of the sample over time with lysozyme enzyme. The nonwoven mats of known weight before degradation (W_i) were immersed in Phosphate Buffer Solution (PBS) (pH 7.4) with and without 10µg/mL lysozyme at room temperature for four weeks which was replaced weekly with freshly prepared lysozyme solution. At various time intervals (1, 2 & 3 weeks, the nonwoven mats were removed and dried overnight. The weight of the dried sponges after degradation (W_d). The percentage of the weight remained after degradation was analyzed using the following formula (Nwe *et al.* 2009),

% Weight remained = $(W_d / W_i) \times 100$ (3.3)

3.2.3.7 Blood compatibility testing – *in vitro* haemolytic activity assay

Haemolysis activity assay was conducted to evaluate the blood compatibility of the nonwoven mats. Fresh and Healthy Human Acid citrate dextrose (ACD) blood (0.25ml) was added onto the wet nonwoven mats which was equilibrated with saline water. After

20 min, 2.0 ml of saline was added onto the nonwoven mats to stop haemolysis and were incubated for 60

min at room temperature. Positive control (PC) and negative control (NC) were obtained by adding 0.25 ml Human ACD into distilled water and saline solution respectively. Incubated mats were centrifuged at 4°C (1500 rpm, 5 min), the supernatant was taken and its absorbance at 545 nm was recorded. The % hemolysis was calculated using the following formula (Liu *et al.* 2015),

$$\text{Haemolysis (\%)} = [(A_{\text{test sample}} - A_{NC} / (A_{PC} - A_{NC}] \times 100 \quad (3.4)$$

3.2.3.8 Cytotoxicity testing – *in vitro* cell viability assay

Viability of the cells was performed using of 3-(4,5-dimethylthiazol-2-yl)-2,5-diphenyl tetrazolium bromide (MTT) assay as per ISO 10993:5 standards. L929 cells used for the assay were grown in regular Dulbeccos Modified Eagles (DME) Medium with 10% fetal bovine serum and antibiotic-antimycotic mix. For MTT assay, the cells were resuspended in fresh culture medium and seeded in control wells or wells containing the different nonwoven mats placed in a 96-well plate. The cells were then placed in a humidified atmosphere of 5% CO_2 for 24 hours. Following incubation, the medium was removed and each well was treated with 0.5 mg/ml of MTT was dissolved in DME Medium and incubated for 2 hours at 37 °C in a humidified atmosphere of 5% CO_2. The yellow MTT is reduced to blue- purple formazan by the mitochondrial dehydrogenase from intact living cells. The resulting blue precipitate was solubilised in Dimethyl sulfoxide (DMSO) and the absorbance was recorded at 570 nm wavelength using a standard plate reader (Gai *et al.* 2020). The percentage cell viability (CV) was calculated using the formula below,

$$CV(\%) = (\text{Absorbance}_{\text{test sample}} / \text{Absorbance}_{\text{control}}) \times 100 \quad (3.5)$$

3.2.3.9 Wound healing Studies

To confirm the wound healing activity of the paramylon functionalized nonwoven mat, one mat is selected with significant property based on the studies done above. Due the low toxicity and high water absorbing capacity, SBM25 was further considered to study the wound healing activity.

A. Dynamic blood coagulation assay

Dynamic blood coagulation assay measures the release of haemoglobin from residual erythrocytes that remained free from entrapment during clot formation. The Assays were performed by recalcification of the anticoagulated whole blood. Briefly, SBM25 nonwoven mat were placed at the bottom of 100 ml beaker and then prewarmed in a water bath at 37 °C for 5 min. Eventually, the steps followed in section 2.2.1.5.A was followed and at the designated termination point, the beaker was removed from the water bath and shaken for 10 min followed by addition of 50 ml distilled water to lyse the free erythrocytes. The absorbance of the supernatants at 540 nm was determined. The absorbance signal derived from the proportion of clot-free erythrocytes are inversely proportional to the size of the clots. The absorbance-time curve was constructed (Liu *et al.* 2014).

B. *In vitro* Paramylon Release Kinetic Assay

This Assay was performed to investigate the release kinetics (burst release vs. sustained release) of paramylon blended with silk fibroin. In order to evaluate the drug release kinetics, SBM25 nonwoven mat were cut into samples of average weight of 5 mg and then the procedures followed for films in section 2.2.1.5.C is followed here also to detect the paramylon release profile at 208 nm in UV spectroscopy. At the beginning of the release test, a

55

linear calibration curve of standard paramylon solution from 5 to 20 μg/ml was created. Paramylon release behaviour in the BPFS sponge sheet were analyzed according to first order model (Cam *et al.* 2020).

C. *In vitro* wound scratch assay

To monitor the cell proliferation via wound scratch assay, the procedures followed in section 2.2.1.5.C is followed here in SBM25 nonwoven mat and the cell proliferation was monitored at different time points: 0 h, 4 h, 18 h, 24 h and images of the migrated cells were taken at all different time points using digital camera (Nikon, Tokoyo, Japan) connected to the inverted phase contrast microscope (Radical instruments, India).

3.3 RESULT AND DISCUSSIONS

3.3.1 Characteristics of Silk fibroin Based Nonwoven Mats

All silk based non-woven mats fabricated were sericin deprived which was confirmed by SEM image of the silk fibers at the time of degumming (Figure 3.1a) and after degumming (Figure 3.1b). The Figure 1b clearly shows the removal of sericin coating (Figure 3.1a) completely with the presence of only core silk fibroin fiber. From the Figure 3.2, the appearance of the silk based non-woven mats are white in colour that forms a network of fibers with the presence of interconnected voids. It was also observed that few fibers are stuck at crossover points, with thin films inlaid amid adjacent fibers that was due to the heterogenous mixture of fibers in acidic solution has good film forming ability due to the partial dissolution of SF fibers in FA-CaCl2 that even helped in covering small interstices among the SF fibers forming a non-woven mat (Pra *et al.* 2006).

Figure 3.1 SEM image of silk Fiber before and after the removal of sericin

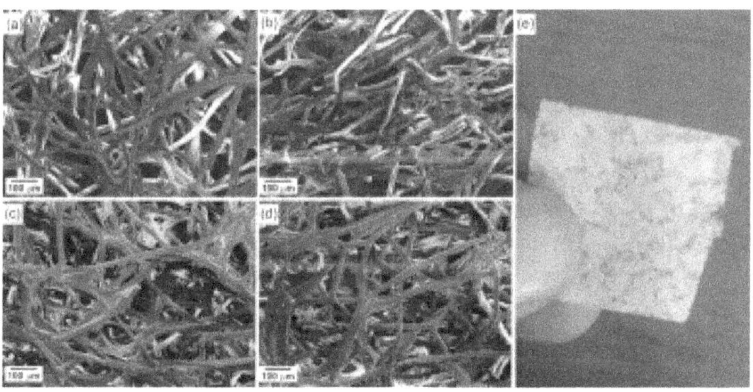

Figure 3.2 Surface morphology of (a) Silk fibroin nonwoven mat, paramylon functionalized silk fibroin nonwoven mat (b) 0.25% (c) 0.5% (d) 1% and Appearance of (e) fabricated nonwoven mat

3.3.2 Molecular Conformation Studies by FTIR Analysis

The molecular conformation of the silk fibroin and paramylon in the silk based nonwoven mats were examined by Fourier Transform Infrared Spectroscopy (FTIR). FTIR is a rapid, nondestructive, time saving method that can detect a range of functional groups and is sensitive to changes in molecular structure. The Figure 3.3 showed the IR absorption bands of nonwoven silk fibroin mat at 1692 cm-1and 1566 cm-1 attributed to the peptide backbone of amide I (C = O stretching) and amide II (C-N stretching) of protein with the β-sheet confirmation (Zhang *et al.* 2012). The bands at 1167 cm-1 and 1109 cm-1 corresponds to the CO-CC stretching and COH bending present in the sugar region (Synytsya & Novak 2014). The bands at 3524 cm-1 corresponds to the O-H bonded with carboxylic acid whereas 2924 cm-1 and 2360 cm-1 corresponds to C-H and N-H stretching (Fareghi-Alamdari *et al.* 2013; Nandiyanyo *et al.* 2019). From the figure, SM and SBM1 peaks follow the same pattern which may imply that the is presence of both SF and BG in the nonwoven mat but there may be no notable chemical interaction in SBM1 mat. But in case of SBM25 and SBM5, there is an interactions in the C-H and N-H stretching group present in them that tends to change the pattern of SM mats on the surface.

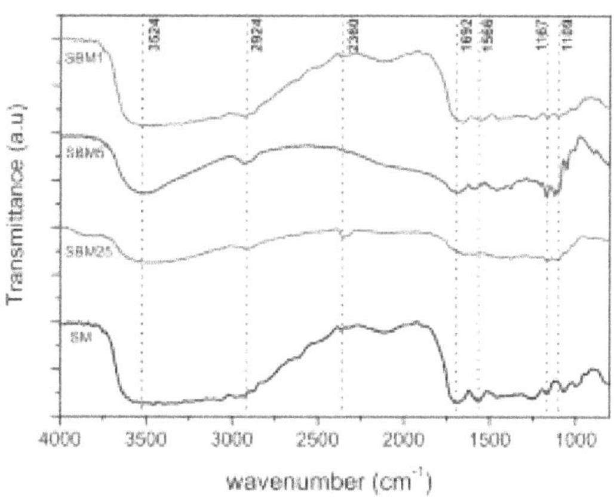

Figure 3.3 FTIR spectrum of Silk fibroin based nonwoven mat

3.3.3 Surface Wettability Testing

The surface wet-ability of the silk fibroin based nonwoven mats functionalized with paramylon was analysed by measuring the contact angle. Wettability of the dressing area is an important factor to maintain the moisture during wound healing process (Fathi *et al.* 2020). From Figure 3.4, it was observed that all the paramylon blended non-woven mats (SBM24, SBM5 & SBM1) showed high to moderate wettability as according to Mitta (2003) the contact angle from 0°to 60° shows high wettability and from 60° to 90° shows moderate wettability. However, SM in the Figure 3.4 shows the contact angle of 124° due to hydrophobic nature of silk fibroin fibers (Murphy & Kaplan 2009).

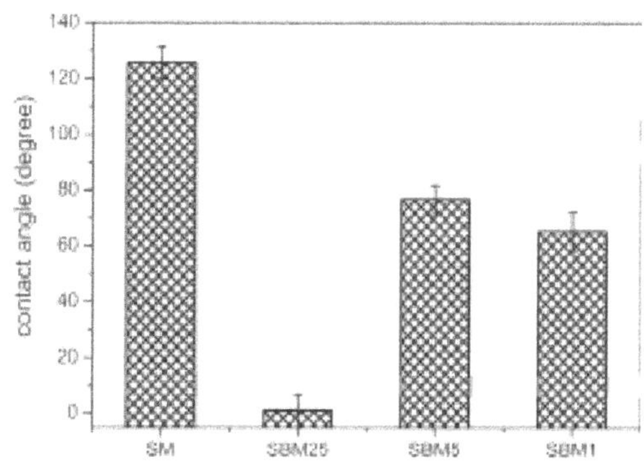

Figure 3.4 Water Contact angle of Silk fibroin based nonwoven mat.

Data represents the mean±standard error (n=3)

3.3.4 Porosity Measurement

The porosity of the silk fibroin based nonwoven mats are shown in Figure 3.5. SM mat had a porosity of 96 %, but the addition of paramylon decreased the porosity to 72 %, 28.571 % and 35.29 % for SBM25, SBM5 and SBM1 which could be due to the plasticizing ability of paramylon that closes the interconnected voids with the films of paramylon (Gan *et al.* 2019). However, the porosity of SBM25 was found to be beneficial for wound dressing as it can help in promoting moisture, oxygen exchange and transfer of nutrients as it has high porosity compared to that of SBM5 and SBM1 (Li *et al.* 2018).

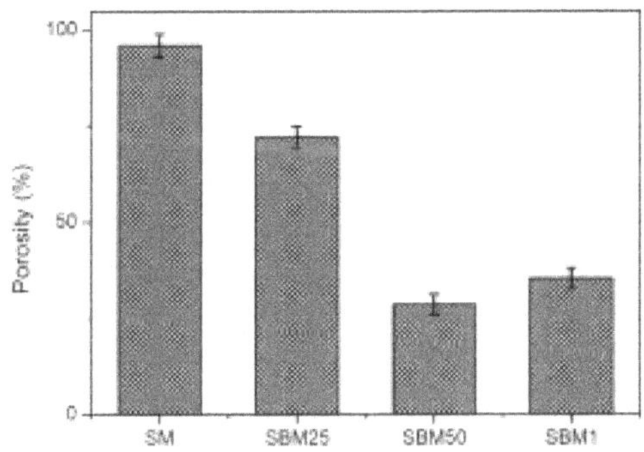

Figure 3.5 Porosity percentage of Silk fibroin based nonwoven mat.

Data represents the mean±standard error (n=3)

3.3.5 Swelling Capacity Studies

An important characteristic of wound dressing is their ability to swell and absorb exudates from the wound site and to maintain a moist atmosphere (Parikh *et al.* 2010). The swelling capacity of the silk based non- woven mats were shown in Figure 3.6. From the Figure 3.6, SBM25 showed maximum of swelling for both PBS and water. However, difference in swelling pattern for SBM1 alone was due to the significant influence of ionic strength of PBS on osmotic pressure (Zeimaran *et al.* 2015).

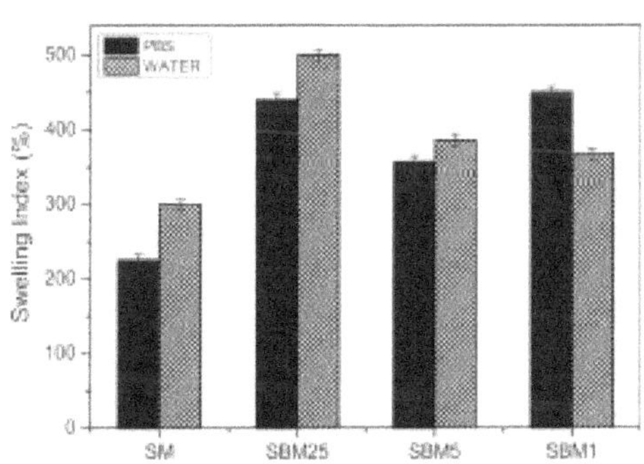

Figure 3.6 Swelling ratio of Silk fibroin based nonwoven mat. Data represents the mean±standard error (n=3)

3.3.6 *In vitro* Enzyme Degradation Studies

Determination of the *in vitro* degradability of the nonwovens allows to predict their stability in aqueous medium and also the degradation capability of the nonwovens in the body enzymes (Santos *et al.* 2020). From the Figure 3.7, it was observed that all the nonwoven mats degrades faster when compared to the other forms as it loses the material stability at the second stage (week 2) itself as the mats got teared off after drying eventhough the weight remained upto 80% of weight upto 2 weeks. Thus for developing wound dressing material, this form can be further layer to 2-ply or 3-ply based on the requirement for wound healing.

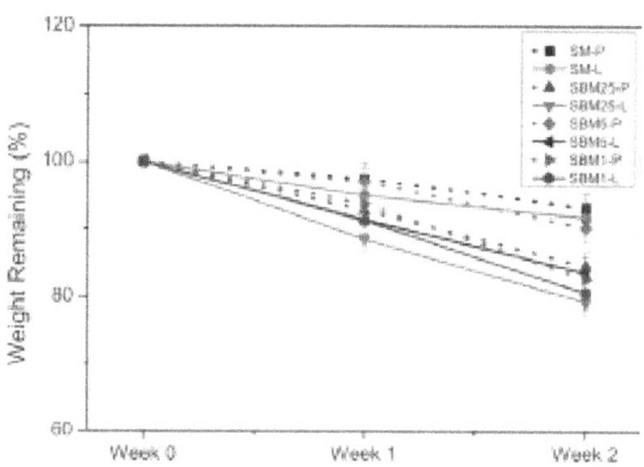

Figure 3.7 Weight remaining percentage of Silk fibroin based non- woven mat in enzymatic(-L) and non-enzymatic degradation environment with PBS(-P). Data represent mean±standard error of the mean(n=3)

3.3.7 Blood Compatibility Testing – *In vitro* Haemolytic Activity Assay

Haemolysis is a very simple and reliable measurement for estimating the blood compatibility and high release of Haemoglobin

(Srivastava & Purwar 2017). The silk fibroin based nonwoven mats were incubated with red blood cells and the effects on erythrocytes were observed. The percentage of Haemolysis were 0.59 %, 0.56 %, 0.72 % and 0.79 % for SM, SBM25, SBM5, SBM1 respectively (Figure 3.8). Therefore, all the nonwoven mats show Haemolysis% below 1% and were considered to be non-haemolytic that do not influence on the wound healing process as free erythrocytes causes nitric oxide inactivation (Zare-Gachi *et al.* 2020).

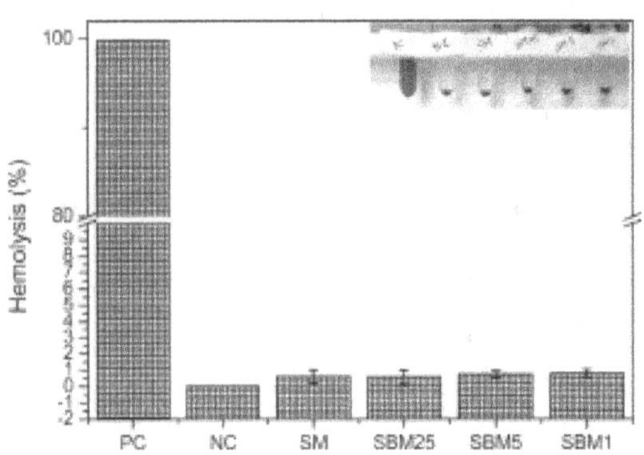

Figure 3.8 Percentage of Haemolytic activity in Silk Fibroin based non- woven mats. Data represent mean±standard error of the mean(n=3)

3.3.8 Cytotoxicity Testing – *In vitro* Cell Viability Assay

The cytotoxicity testing is the biological evaluation and screening that use the tissue cells *in vitro* to observe the cell viability. It is also a basic biocompatibility assay (Singh *et al.* 2017). The cell viability percentage of L929 cells on nonwoven mats are shown in the Figure 3.9. It was observed that SBM25 showed viability entirely comparable to the control plate which has very good biocompatible, whereas other nonwoven mats showed almost 80% of cell viability.

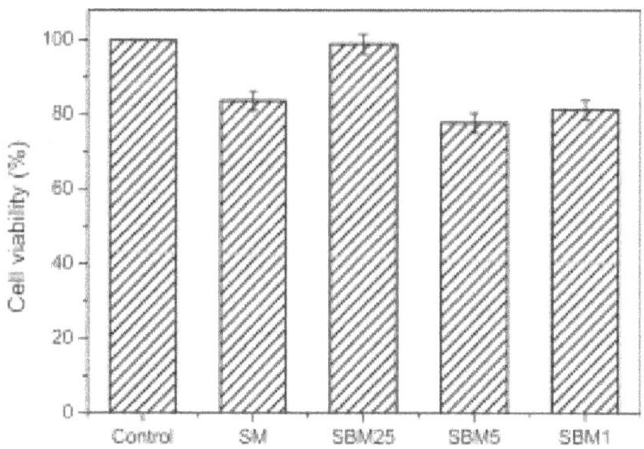

Figure 3.9 Effect of Silk fibroin based nonwoven mats on L929 cell viability. MTT assay of L929 cells for 24 h. hours. Data represent mean±standard error of the mean(n=3)

3.3.9 Dynamic Blood Clotting Assay

The dynamic blood clotting time is an *in vitro* test that measures the degree of activation of intrinsic coagulation factors when surfaces interact with blood. Activation is measured by the absorbance-time curve for release of Haemoglobin from erythrocytes that avoid being entrapped during clot formation (Liu *et al.* 2014). The curve (Figure 3.10) implies that the nonwoven mat shows similar pattern as on control and hence it does not show any adverse effect in the coagulation pathways. The clot sizes appear to be higher in SBM25 nonwoven mat, since the magnitude of absorbance is inversely related to clot size.

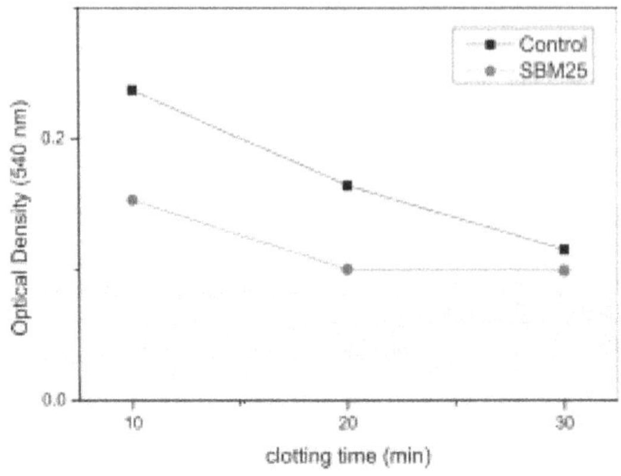

Figure 3.10 Intrinsic coagulation factors activation measurement curve.

Data represent mean±standard error of the mean(n=3)

3.3.10 *In vitro* Paramylon release Kinetic Assay

In vitro drug release kinetic assays for SBM25 nonwoven mat was performed according to the first-order model. Primarily, the UV spectra were obtained with the concentration range of paramylon from 5 to 20 μg/ml and a linear standard calibration curve was drawn from paramylon absorption values (R2 = 0.99127) obtained for the quantitative determination from UV absorbance measured at 208 nm. The *in vitro* release studies were performed over a period of 48 hours for SBM25 nonwoven mat in order to determine the release kinetics of blended bioactive paramylon. The nonwoven mat showed a burst release in approximately 30-60 minutes and maintained a sustained release after the burst release (Figure 3.11). These results are consistent with previous studies in which the burst release reaches high concentration of bioactive molecule and a sustained release afterwards may help in wound closure (Cam *et al.* 2020).

64

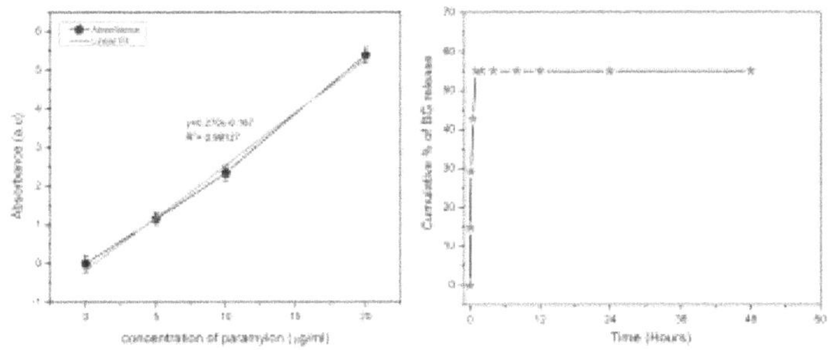

Figure 3.11 *In vitro* drug release profile of Silk Fibroin based nonwoven mat: (a) Paramylon calibration curve (b) Paramylon release profile of SBM25 nonwoven mat according to first-order model. Data represent mean±standard error of the mean(n=3)

3.3.11 *In vitro* Wound Scratch Assay

Cell migration is an essential process in many physiological and pathological events such as wound healing, embryonic development and cancer metastasis. The basic procedure to study cell migration involves (i) creation of a cell monolayer, (ii) the production of a pseudo-wound in the confluent layer of cells, (iii) the capture of images at different time intervals until wound closure is reached, and (iv) the analysis of the image sequence in order to quantify the migration of the chosen cells (Cappiello *et al.* 2018). The *in vitro* scratch assay is an easy and economic method to mimic the cell migration during wound healing *in vivo*. L929 cells were incubated in presence and absence of the SBM25 nonwoven mat and images were captured at 0, 4, 18 and 24 h at various concentrations SBM25 nonwoven mat (25, 50, 75 and 100 µg). From the Figure 3.12, it is clearly observed that the cell migration has started at the 4 hours after incubation in almost all the concentration and has reached upto 92-99 % within 24 h. Thus, confirming the wound healing capacity of SBM25 nonwoven mat on L929 Fibroblast cells.

65

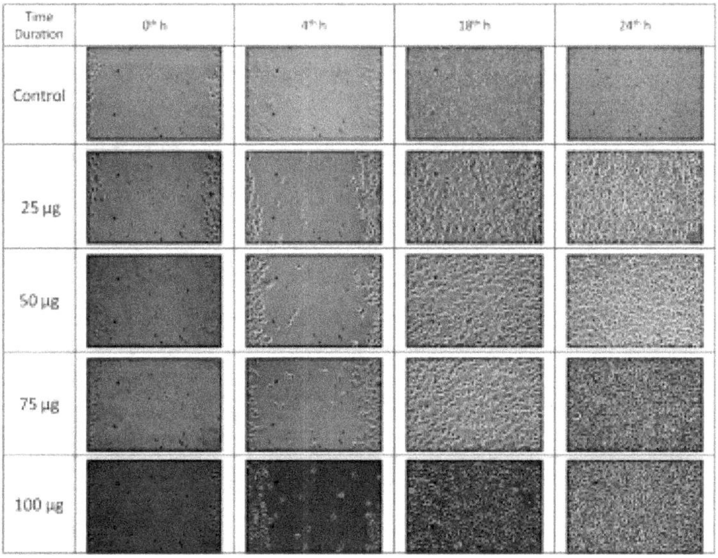

Figure 3.12 Scratch wound assay using L929 fibroblasts incubated with SBM25 nonwoven mat (25, 50, 75 & µg) at 0, 4, 18, 24 hours

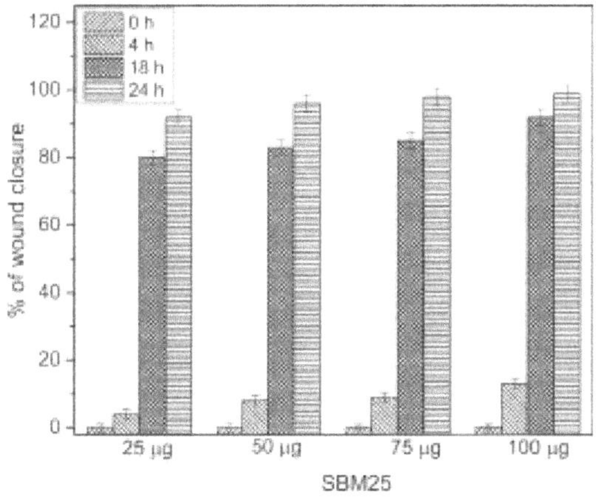

Figure 3.13 Percentage of wound closure at 0, 4, 18, 24 hours. Data represent mean±standard error of the mean(n=3)

3.4 CONCLUSION

To solve the challenges of chronic wound management and to help the patients and health care systems during this pandemic situation, this bioactive nonwoven silk fibroin mat is blended with paramylon for self- treating the chronic wounds. The paramylon blended Silk fibroin nonwoven mats were successfully prepared in 3 different ratios such as 0.25%, 0.5%, 1% paramylon and its various parameters like wettability, swelling, porosity, degradation, Hemocompatibility and cytotoxicity studies were done. From the studies, it was observed that the Silk fibroin nonwoven mat with 0.25% paramylon showed good wettability, porosity with high swelling capacity, good Hemocompatibility and very good biocompatibility. It was also confirmed through the *in vitro* wound scratch assay, that SBM25 has a good wound healing capacity as the cell migration increases with increase in concentration of the SBM25 and has produced up to 92-99% wound closure in 24 h. Thus, SBM25 can further be optimized and analysed to develop an adhesive self-care bandage for chronic wound patients.

CHAPTER 4

FABRICATION AND EVALUATION OF ALUM LOADED SILK FIBROIN/PARAMYLON SUPPORTED ON CELLULOSE SPONGE FOR HEAVILY EXUDING CHRONIC WOUNDS

This chapter deals with the Preparation of alum loaded Silk fibroin and Paramylon blends coated cellulose sponge for impaired wound healing. A detailed procedures about the extraction of silk fibroin and fabrication of cellulose sponge supported Silk fibroin and Paramylon blends were presented along with the various physical and biological characterization related to the wound dressings.

4.1 INTRODUCTION

Highly exuding wounds causes great distress to patients. Developing an efficient and cost-effective management of excessive wound exudates continues to be still challenging (Benbow & Stevens 2013). These exudates play a vital role in delaying the healing process by making the wound chronic. A Chronic wound may be defined as a wound that failed to proceed through a timely and orderly healing process (Frykberg & Banks 2015). These non-healing wounds can be enervating for the affected individual by causing pain, loss of mobility, depression, anxiety, chronic morbidity or even death and can place a financial burden on healthcare systems. Acute wounds normally heal in sequential and timely stages like hemostasis, inflammation, proliferation and remodeling (Landen *et al.* 2016). But the chronic wounds

68

healing remains halted in one of these stage thereby developing a non-healing wound for many reasons (Iqbal *et al.* 2017). However, uncontrolled matrix metalloproteinases (MMPs) are one of the major cause for the chronicity of non-healing ulcers. The preponderance of MMPs in the wound fluid/exudates is characterized as spellbound in an ongoing inflammatory process with a continual breakdown of Extracellular matrix (ECM) and its components (McCarty & Percival 2013). This chronic inflammatory wound exudate consists of polymorphonuclear leukocytes, macrophages and neutrophils that secrets proteases and multiple cytokines (Krishnaswamy *et al.* 2017). When comparing wound exudate to wound tissue, wound fluid contains higher level of MMP-9 which predict poor healing in chronic wounds like pressure ulcers and diabetic foot ulcers. MMP-9 is not expressed in normal skin but it is necessary for normal progression of wound healing. Keratinocyte migration, re-epithelialization and angiogenesis are dependent on MMP-9 but in excess MMP-9 delays wound healing in the background causing prolonged inflammation. According to the highly advanced technology healing enhancements such as growth factors and bioengineering skin was found to have least performance if the excess exudates are not controlled properly (Falanga 2000). Thus, to control the level of MMPs and to reduce the chronicity, high wound exudate in the non-healing wounds that cause infection must be regulated by providing highly absorbable wound dressing material creating a moist wound environment (Widgerow 2011).

A perfect wound dressing must help in maintaining the moisture of the wound, allow oxygen exchange, absorb wound exudates, accelerate re- epithelialization for wound closure preventing infections, reduce pain and healing time. All these requirements renowned focus on creating an instructive wound dressing material.

The natural abundance, low cost, biodegradability, eco-friendly nature and hydrophilicity have attracted increasing interest on Cellulose (Naomi & Fauzi 2020). The cellulose and its derivatives, have been widely used in healthcare and biological applications such as scaffolds, drug diagnosis and orthopedics (Dumanli 2017). Due to its ability to retain moisture, cellulose gauze-based bandages have been used for a long time in clinical practice but it does not possess any healing property. The macro and micro pores of the cellulose sponge help in inducing gas exchange, absorbing excess wound exudates and maintaining moist environment (Pei *et al.* 2015). The cellulose is thereby used as the porous matrix for supporting bioactive materials for healing high drainage chronic wounds by maintaining moist wound environment. Besides with cellulose, to accelerate wound healing activity by reducing infection and healing time, combination of silk fibroin and paramylon are used for developing a bioactive instructive self-care wound dressing for high exudate chronic wounds.

Fibroin, a natural polymer extracted from silk cocoon has been well-known for its diverse applications in the biomedical field such as wound healing (Gil *et al.* 2013), tissue regeneration (Wang *et al.* 2012) and drug delivery (Guziewicsz *et al.* 2011) due its strong mechanical strength, controlled biodegradability, non-cytotoxic, low-antigenicity and non- inflammatory nature (Mauney *et al.* 2007). In specific silk fibroin extracted from Bombyx mori has gained more attention as it possesses unique biological properties like biocompatibility, water and oxygen uptake, low immunogenicity and robust mechanical behaviour (Farokhi *et al.* 2018). The other polymer is a multifunctional polysaccharide called paramylon. Paramylon (BG) is a type of β-glucan comprised of β-linked D-glucose molecules extracted from *Euglena gracilis*. The immunomodulatory properties of β-glucan have been known for decades and it's also available as a dietary supplement commercially (Vannucci *et al.* 2017). In which, paramylon can be produced in high quantity through biotechnological process which is economical also. This paramylon possesses immune-potenting activity (Nakashima *et al.* 2017), anti-atopic dermatitis effects (Sugiyama *et al.* 2010), hepato-protective effect (Sugiyama *et al.* 2009), anti-microbial activity (Sakagami *et al.* 1991) and also biocompatible (Shibakami *et al.* 2013). Also, According to Arthe *et al.* (2020), the silk fibroin and paramylon in the ratio of 75:25 showed better property with highest Haemocompatibility, biocompatibility possessing more than 90% of cell viability against HEK 293T cells. Thus, to develop an instructive self-care dressing, both cellulose and SF/BG role is important for high exudate wounds. To make them adhere each other without changing the nature of the cellulose sponge, natural mordant that is most frequently used by the dyers for protein, plant fibers and fabrics were used. Alum is one such natural mordant that improves light and wash fastness of all- natural dyes and keeps colours clear. It is also less expensive and non- toxic (Bnyan *et al.* 2014). Alum is also used as a drug in Ayurveda and Unani medicinal system (Ali *et al.* 2017), food processing and preservation (Efiuwevwere & Amadi), antibacterial agent (Al-Talib *et al.* 2016), cosmetics (Alzomor *et al.* 2014) and textile industry (Mozaffari & Maleki 2018) and hence it was loaded in FG 75:25 solution and fixed on cellulose sponge support.

4.2 EXPERIMENTAL SECTION

4.2.1 Materials

Bombyx mori silk cocoons were kindly supplied by the Department of studies in Sericulture sciences, University of Mysore. β(1-3)-D-glucan used were derived from *Euglena gracilis* and was purchased from Sigma-Aldrich Chemical Co.(St. Louis, MO.) and used without further treatment. Cosmetic grade cellulose sponge was purchased commercially and used without any further treatment. All other chemicals used were analytical grade obtained commercially. L929 cell line (murine fibroblast) was purchased from National Centre for Cell Science (NCCS), Pune.

4.2.2 Methodology

4.2.2.1 Extraction of Silk fibroin from *Bombyx Mori* Cocoon (Bivoltine)

According to Rockwood *et al.* (2011), The procedures or steps followed in section 2.2.1.2 was done for the extraction of silk fibroin from Bombyx Mori Cocoon

4.2.2.2 Preparation of alum loaded silk fibroin / paramylon blends coated on cellulose sponge

The blends of silk fibroin and paramylon was prepared in the ratio 75:25 to form 1% solution by weight with 98% formic acid and named as FG 75:25. FG 75:25 blends prepared was mixed with 1% Potash Alum (P) solution in the ratio 80:20 (AP) and coated on the cellulose using saturation coating method which was named as CAP. FG 75:25 blends alone coated on cellulose sponge was named as CA and 1% Alum on Cellulose sponge as CP and Cellulose Sponge as a blank named as C. The saturation coating method is also known as dip coating or impregnation and it is a simple process of uniform coating onto the flat surface where the substrate is immersed in the solution which can be removed by evaporation that results in unique coating (Gulrajani 2013).

4.2.3 Characterization Studies

4.2.3.1 SEM-EDX analysis

The Surface morphologies and elemental composition of the sponges was examined using Carl Zeiss MA15/EVO 18 SEM coupled with energy-dispersive X-ray (Oxford Instruments NanoAnalysis INCA Energy 250 Microanalysis System (EDX). About 0.5 cm of the dry sponges were cut and sputter coated with gold to enhance the surface conductivity. EDX spectra were measured with an energy of the primary electron beam of 20keV.

4.2.3.2 Fourier transform infrared (FTIR) spectroscopy analysis

Secondary structure and conformational changes were confirmed by FTIR spectroscopy with air as the reference in the spectral range of 4000 - 400 cm^{-1} at 4 cm^{-1} spectral resolution and 32 scans using JASCO International Co./Japan Fourier transform infrared Spectrometer Model FTIR-6300 instrument.

4.2.3.3 Porosity measurement by solvent replacement method

Dry sponges were immersed overnight in absolute ethanol and weighed after excess ethanol on the surface was blotted. The porosity was calculated from the below formula (Nanda *et al.* 2013)

Porosity = $[(M_2 - M_1)/\rho V] \times 100$ (4.1)

4.2.3.4 Swelling capacity testing by gravimetric method

The dry sponges of known weight (W_d) were immersed in both water and PBS of pH 7.4 at room temperature for at least 24 hours. After which the water or PBS was removed from the sponges by gently dabbing onto a tissue paper and the swollen weight (W_s) was recorded. Then, the swelling index (SI) was calculated with the below formula (Zeimaran *et al.* 2015)

$SI = \{(W_s - W_d) / W_d\} \times 100$ (4.2)

4.2.3.5 *In vitro* Enzyme Degradation Studies

Degradation of Cellulose supported sponges were determined as percentage of weight remained after incubation in lysozyme enzyme solution for about 4 weeks. The sponges with known dry weight before degradation(W_0) were immersed in Phosphate Buffer Solution (PBS) (pH 7.4) with and without 10µg/mL lysozyme at room temperature for four weeks which was refreshed weekly. At various time intervals (1, 2 & 3 weeks, the sponges were removed and dried overnight. The weight of dried sponges after degradation (W_t). The percentage of the weight remained after degradation was analyzed using the formula below (Luangbudnark *et al.* 2012).

% Weight remained = (W_t / W_0) x 100 (4.3)

4.2.3.6 Blood compatibility testing – *in vitro* haemolytic activity assay

Haemolysis activity assay was conducted to evaluate the blood compatibility of the sponges. In this experiment, dry test sponge was equilibrated in saline water (0.9% NaCl solution) for 24 hrs at 37°C. Healthy Human ACD blood (0.25ml) was added onto the wet sponges. After 20 min,

2.0 ml of saline was added onto the sponges to stop Haemolysis and the test samples were incubated for 60 min at room temperature. Positive and negative controls were obtained by adding 0.25 ml Human ACD into distilled water and saline solution respectively. Incubated sponges were centrifuged at 4°C (1500 rpm, 5 min), the supernatant was taken and its absorbance at 545 nm was recorded. The % Haemolysis was calculated using the following equation (Kim *et al.* 2014),

Haemolysis (%) = [($A_{test\ sample}$ – $A_{(-)\ control}$) / ($A_{(+)\ control}$ – $A_{(-)\ control}$)] x 100 (4.4)

4.2.3.7 Cytotoxicity testing – *in vitro* cell viability assay

Viability of the cells was performed using MTT assay. For this assay, the steps involved in section 3.2.3.8 was repeated on cellulose based sponges against L929 cells and the resulting blue precipitate was solubilised in Dimethyl sulfoxide (DMSO) and the absorbance was recorded at 570 nm wavelength using a standard plate reader (Wiegand *et al.* 2019). The percentage cell viability (CV) was calculated using the formula below:

$$CV(\%) = (\text{Absorbance of test sample} / \text{Absorbance of control}) \times 100 \quad (4.5)$$

4.2.3.8 Wound healing Studies

To confirm the biological activity of the Alum loaded FG 75:25 blends coated Cellulose sponge, CAP was further considered to study the wound healing activity as it showed less toxicity when compared to other sponges and also it has a moderate swelling and degradation property as it has a predominant coating of FG75:25.

A. Dynamic Blood Coagulation Assay

Dynamic blood coagulation assay measures the release of haemoglobin from residual erythrocytes that remained free from entrapment during clot formation. The Assays were performed by recalcification of the anticoagulated whole blood. Briefly, CAP Sponge were placed at the bottom of 100 ml beaker and then prewarmed in a water bath at 37 °C for 5 min. Eventually, the steps followed in section 2.2.1.5.A was followed and at the designated termination point, the beaker was removed from the water bath and shaken for 10 min followed by addition of 50 ml distilled water to lyse the free erythrocytes. The absorbance of the supernatants at 540 nm was determined. The absorbance signal derived from the proportion of clot-free erythrocytes are inversely proportional to the size of the clots. The absorbance-time curve was constructed (Liu *et al.* 2014).

B. *In vitro* Paramylon Release Kinetic Assay

This Assay was performed to investigate the release kinetics (burst release vs. sustained release) of paramylon blended with silk fibroin. In order to evaluate the drug release kinetics, CAP sponge were cut into samples of average weight of 5 mg and then the procedures followed for films in section

2.2.1.5.C is followed here also to detect the paramylon release profile at 208 nm in UV spectroscopy. At the beginning of the release test, a linear calibration curve of standard paramylon solution from 5 to 20 μg/ml was created. Paramylon release behaviour in the BPFS sponge sheet were analyzed according to first order model (Cam *et al.* 2020).

C. *In vitro* wound scratch assay

To monitor the cell proliferation via wound scratch assay, the procedures followed in section 2.2.1.5.C is followed here in CAP sponge and the cell proliferation was monitored at different time points: 0 h, 4 h, 18 h, 24 h and images of the migrated cells were taken at all different time points using digital camera (Nikon, Tokoyo, Japan) connected to the inverted phase contrast microscope (Radical instruments, India).

4.3 RESULT AND DISCUSSIONS

4.3.1 Characteristics of Cellulose Supported Sponges

All the cellulose supported sponges were prepared using saturation coating method had a porous, spongy appearance and white in colour as shown in Figure 4.1. The thickness of the sponges was around 0.5cm and the water-based dye drops shown confirms the hydrophobic nature of A on the surface of CA and CAP sponges.

The SEM microstructure of the cellulose sponge and cellulose supported sponges fabricated are shown in the Figure

4.2. The porous nature of the cellulose sponge was very clear in the sponges C and CP compared to that of Cellulose with SF/BF blends as the pores were coated with the blends firmly on the surface alone. Also, it is clear that the alum loaded FG 75:25 coated cellulose sponge in coated efficiently as almost all the pores were fully covered with the coating. Table 4.1 consist of the elemental composition of the sponges revealing the presence of various elements in which magnesium and chlorine ions show large changes in CAP compared to that of other sponges. Studies revealed that the natural sponge obtained from plant source is softened by using Magnesium Chloride as the softening agent (Chengyong, 2010) thus making it more convincing that FG 75:25 coating is predominant in the surface of CAP which reduces the percentage of magnesium chloride.

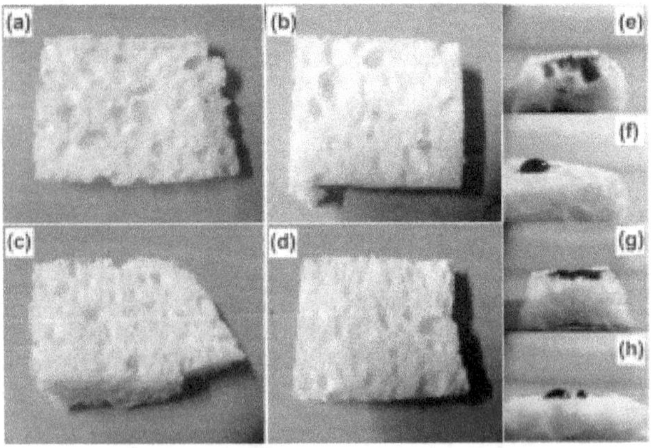

Figure 4.1 Porous nature of the sponges (a) C (b) CA (c) CP (d) CAP, (e)-(g) hydrophillic surface of C-CP, (f)-(h) hydrophobic surface of CA-CAP

77

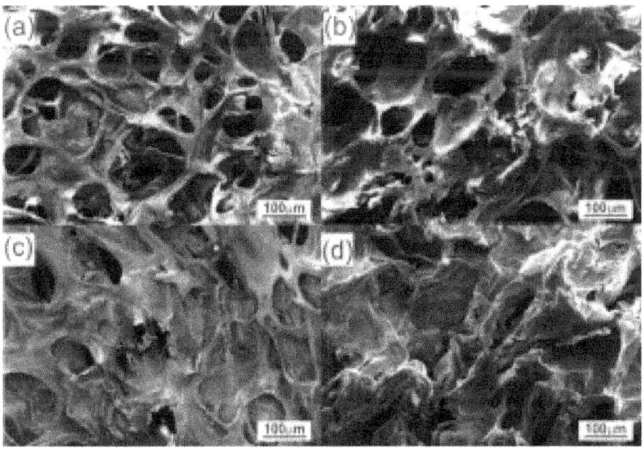

Figure 4.2 SEM images of (a) cellulose-C and cellulose sponges supported Alum-CP (b), FG 75:25 blends-CA(c), and both Alum and FG 75:25 blends-CAP (d)

Table 4.1 EDX Elemental Analysis of all sponges

Sample Code	Elements Weight Percentage (%)

	C	O	N	Mg	Cl	Al	S	K
C	41.28	48.60	-	3.74	6.38	-	-	-
CA	36.46	49.27	1.48	4.00	8.79	-	-	-
CP	40.34	49.96	-	2.99	5.35	0.65	0.51	0.20
CAP	41.35	48.59	6.66	0.25	0.58	0.95	1.60	0.02

4.3.2 Molecular Conformation Studies by FTIR Analysis

The FTIR spectra of pure cellulose sponge, FG 75:25 blends and FG 75:25 blends supported cellulose sponge with and without alum were shown in the Figure 4.3. The Characteristic profile of typical cellulose was detected in the regions of 3700-3100 cm-1 (hydrogen bonding in cellulose); C- H Stretching at 3000-2850 cm-1; C-H bending (Scissoring & wagging) at 1500-1300 cm-1 and the region between 1200-950 cm-1 corresponds to the C-

O-C stretching vibration in the C peak (Abidi *et al.* 2014). Due to the presence of lignin impurities, the peak was produced at 1595 cm-1 that corresponds to the C=C Stretching of the aromatic ring in lignin (Esteves *et al.* 2013). The Characteristic peak of SF was found at 1648 cm-1, 1537 cm-1 and 1242 cm-1 corresponds to the amide I, amide II and amide III Linkage of protein with both β-sheets and random coil structure (Arthe *et al.* 2020). The peak at 1441 cm-1 corresponds to the in-plane ring deformation including CH- OH bending modes. The paramylon bands are at 1147 and 1043 cm-1 corresponds to the CO-CC stretching and COH bending mode present in sugar region. The peaks of CA and CAP confirm the coating of FG 75:25 on cellulose by reduced intensities. The peak shift near 1147 cm-1 in AP is attributed to the alum which may be due to the S-O vibrations of sulfate and other peaks of alum may include Al-OH-Al stretching mixed with OH vibrations and O-H bending and unsymmetrical stretching of H_2O molecule co-ordinate with Al (Lal & Garg 2017). According to Wenk *et al.* 2010, the sulfonated silk fibroin can be represented as a beneficial biomaterial for efficient loading of growth factors and the OH which may also help in healing process.

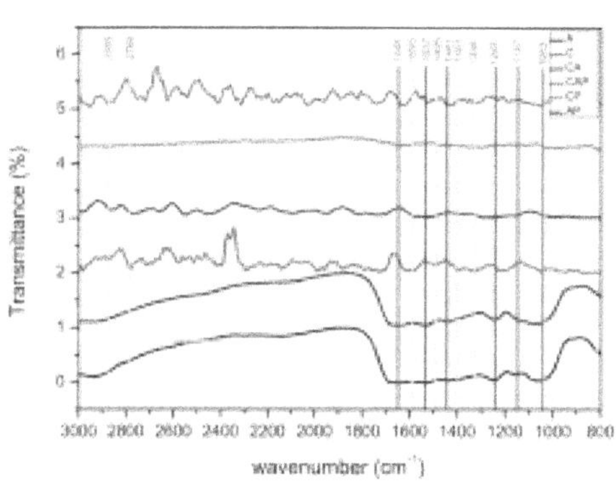

Figure 4.3 FTIR spectrum of Film (A, AP), Cellulose Sponge (C), Cellulose sponge supported Alum (CP), FG 75:25

blends (CA) and with both Alum & FG 75:25 blends (CAP)

4.2.3 Porosity Measurement of the Sponges

In general, the porous network structure in wound dressing can absorb exudates and maintain the suitable moisture environment, facilitate oxygen exchange and controls vapour evaporation (Pei *et al.* 2015 ; Hu *et al.* 2019). From the Figure 4.4, It was confirmed that the cellulose sponge provides porous support to FG 75:25 blends and Potash Alum for making it more absorbing to maintain gaseous exchange as the porosity percentages of all cellulose supported sponges was found to be more than 90 %.

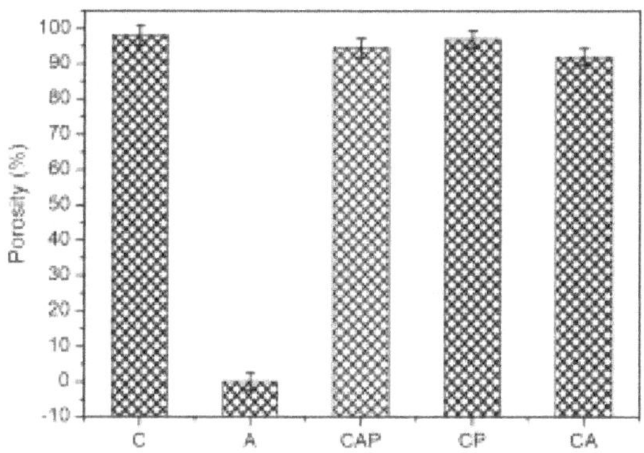

Figure 4.4 Porosity of Film (A), Cellulose Sponge (C), Cellulose sponge supported Alum (CP), FG 75:25 blends (CA) and with both Alum & FG 75:25 blends (CAP). Each bar represents mean±standard error (n=3)

4.2.4 Swelling Capacity Studies

The Swelling capacity is an important characteristic for the wound dressing material especially for high exudates wounds (El-Kased *et al.* 2017). Penetration of solvent leads to swelling of the sponges due to the diffusion of water molecules. Figure 4.5 showed the fluid holding capacity of the cellulose supported sponges. Due to its macro-porous nature of cellulose, the water molecules diffuse into the cellulose very well compared to A, CA, CP and

CAP. Meanwhile, the difference in the swelling index in water and PBS is caused by a higher ion concentration of PBS that leads to low osmotic pressure resulting in a lower absorption capacity (Kusumastuti *et al.* 2019). It also confirms that there is an increase in swelling ratio of the cellulose supported sponge compared to that of FG 75:25 blends, which makes it evident to absorb the excess exudate from the wounds.

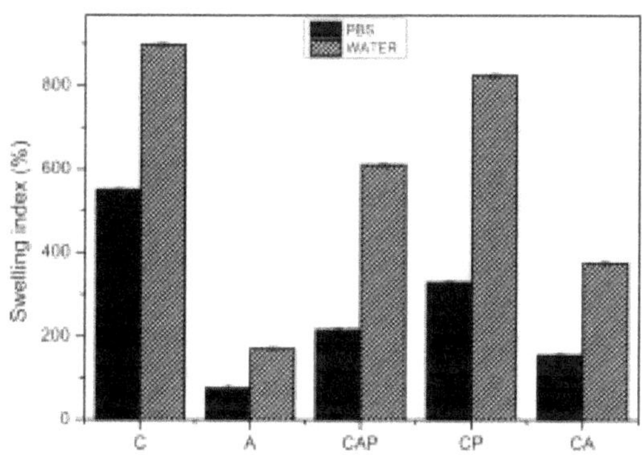

Figure 4.5 Swelling capacity of Film (A), Cellulose Sponge (C), Cellulose sponge supported Alum (CP), FG 75:25 blends (CA) and with both Alum & FG 75:25 blends (CAP) . Each bar represents mean±standard error (n=3)

4.3.5 *In vitro* Enzyme Degradation Studies

The biodegradation of the wound dressing sponge was studied by lysozyme hydrolysis. The residual weight of the sponges after immersion were inversely proportional to the degradation rate (Huang *et al.* 2010). From Figure 4.6, it has been observed that all the sponges maintained nearly 50% of its original weight for about 3 weeks and also the sudden decrease in weight of the sponge in the first seven days may be due to the shrinkage of original macro-porous nature of the sponges as it almost maintained its weight after seven days. Figure 4.6 and Figure 4.5, it was observed that the quicker it absorbs may leads to more degradation by shrinkage of macro-porous nature after attaining maximum fluid holding capacity.

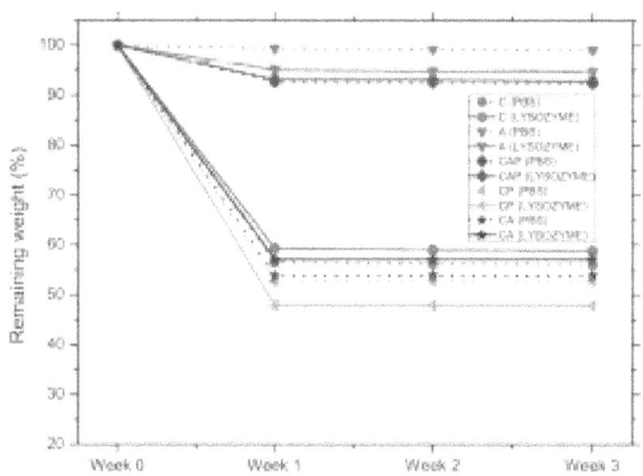

Figure 4.6 Weight remaining percentage of FG 75:25 (A) Cellulose Sponge (C), Cellulose sponge supported Alum (CP), FG 75:25 blends (CA) and with both Alum & FG 75:25 blends (CAP) in enzymatic and non-enzymatic degradation environment. Data represent mean±standard error of the mean(n=3)

4.3.6 Blood Compatibility Testing – *In vitro* Haemolytic Activity Assay

Haemolysis analysis has been done to evaluate the adverse effect of the sponges on the blood cells. During haemolysis, red blood cells are destroyed and the free Hb can be detected in the surrounding environment and influence the wound healing process by causing Nitric oxide inactivation and disturb fibroblast proliferation and collagen remodelling (Zare-Gachi *et al.* 2020). From Figure 4.7, it was observed that all the sponges obtained values closer to that of the negative control which confirms that the sponges do not show any traces of disintegration of blood cells and also, they show good blood compatibility.

82

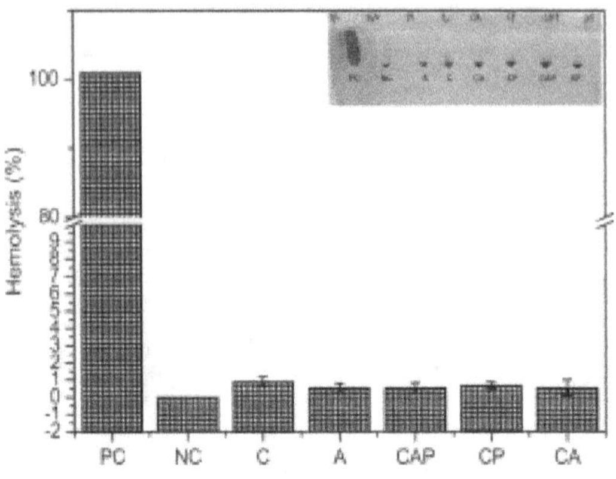

Figure 4.7 Percentage of Haemolytic activity in FG 75:25 (A), Cellulose Sponge (C), Cellulose sponge supported Alum (CP), FG 75:25 blends (CA) and with both Alum & FG 75:25 blends (CAP). Data represent mean±standard error of the mean(n=3)

4.3.7 Cytotoxicity Testing – *In vitro* Cell Viability Assay

The toxicity of the sponges was investigated based on the effect of blank cellulose sponge with and without A, P or AP coating on the viability of L929 cells after the incubation of 24 hours by MTT reduction assay. Cytotoxicity is inversely proportional to cell viability rated as severe (<30%), moderate (30-60%), mild (60-90%) or non-cytotoxic (>90%) based on the activity relative to values obtained for the controls (Dahl *et al.* 2006). From the Figure 4.8, it was observed that all cellulose sponges except CAP possess moderate cytotoxicity which may be due to the presence of lower Magnesium in CAP (Table 4.1) as the high concentration of Magnesium can induce even death of the L929 cells (Zhen *et al.* 2015).

83

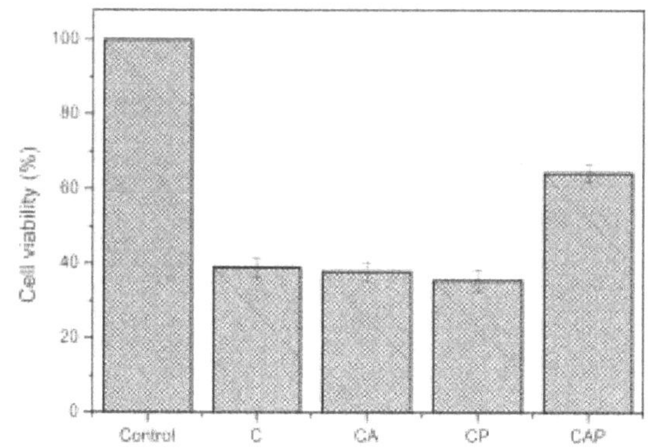

Figure 4.8 Effect of Cellulose Sponge (C), Cellulose sponge supported Alum (CP), FG 75:25 blends (CA) and with both Alum & FG 75:25 blends (CAP) on L929 cell viability. MTT assay of L929 cells for 24 h. hours. Data represent mean±standard error of the mean(n=3)

4.3.8 Dynamic Blood Coagulation Assay

The dynamic blood clotting time is an *in vitro* test that measures the degree of activation of intrinsic coagulation factors when surfaces interact with blood. Activation is measured by the absorbance-time curve for release of Haemoglobin from erythrocytes that avoid being entrapped during clot formation (Liu *et al.* 2014). The curve (Figure 4.9) implies that the CAP sponge shows similar pattern as on control and hence it does not affect the coagulation pathways. The clot sizes appear to be higher in CAP sponge, since the magnitude of absorbance is inversely related to clot size due to the presence of Potash Alum which is the well-known drug in Ayurvedic and unani medicine to stop the bleeding in wounds.

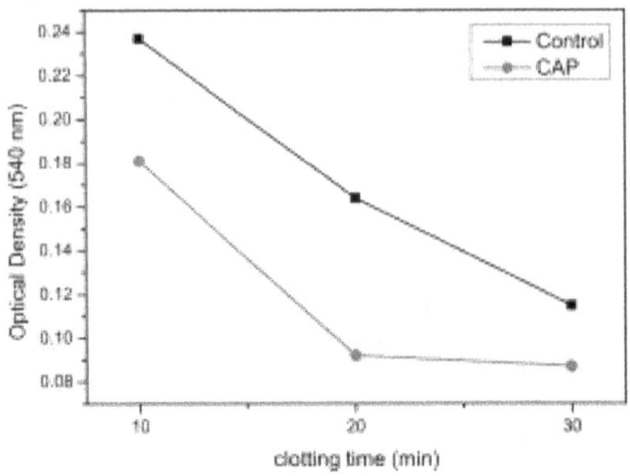

Figure 4.9 Intrinsic coagulation factors activation measurement curve.

Data represent mean±standard error of the mean(n=3)

4.3.9 *In vitro* Paramylon release Kinetic Assay

In vitro drug release kinetic assays for CAP sponge were performed according to the first-order model. Primarily, the UV spectra were obtained with the concentration range of paramylon from 5 to 20 µg/ml and a linear standard calibration curve was drawn from paramylon absorption values (R2 = 0.99127) obtained for the quantitative determination from UV absorbance measured at 208 nm. The *in vitro* release studies were performed over a period of 48 hours for CAP sponge in order to determine the release kinetics of blended bioactive paramylon. The sponge showed a burst release in approximately 8 hours and maintained a sustained release after 12 hours (Figure 4.10). These results are consistent with previous studies in which the burst release reaches high concentration of bioactive molecule and a sustained release afterwards may help in wound closure (Cam *et al.* 2020).

85

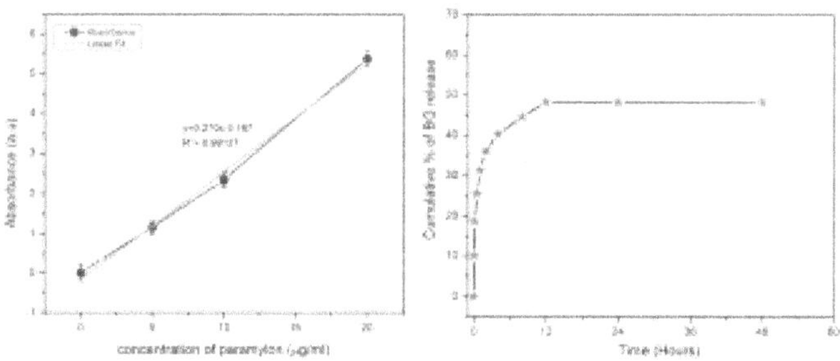

Figure 4.10 *In vitro* drug release profile of Cellulose supported sponge:

(a) Paramylon calibration curve (b) Paramylon release profile of CAP Sponge according to first-order model. Data represent mean±standard error of the mean(n=3)

4.3.10 *In vitro* Wound Scratch Assay

Cell migration is an essential process in many physiological and pathological events such as wound healing, embryonic development and cancer metastasis. The basic procedure to study cell migration involves (i) creation of a cell monolayer, (ii) the production of a pseudo-wound in the confluent layer of cells, (iii) the capture of images at different time intervals until wound closure is reached, and (iv) the analysis of the image sequence in order to quantify the migration of the chosen cells (Cappiello *et al.* 2018). The *in vitro* scratch assay is an easy and economic method to mimic the cell migration during wound healing *in vivo*. L929 cells were incubated in presence and absence of the CAP sponge and images were captured at 0, 4, 18 and 24 h at various concentrations of CAP sponge (25, 50, 75 and 100 µg). From the Figure 4.11, it is clearly observed that the cell migration has started at the 4 hours after incubation in almost all the concentration and has reached up to 97-99 % within 24 h. Thus, confirming the wound healing capacity of CAP sponge on L929 Fibroblast cells.

86

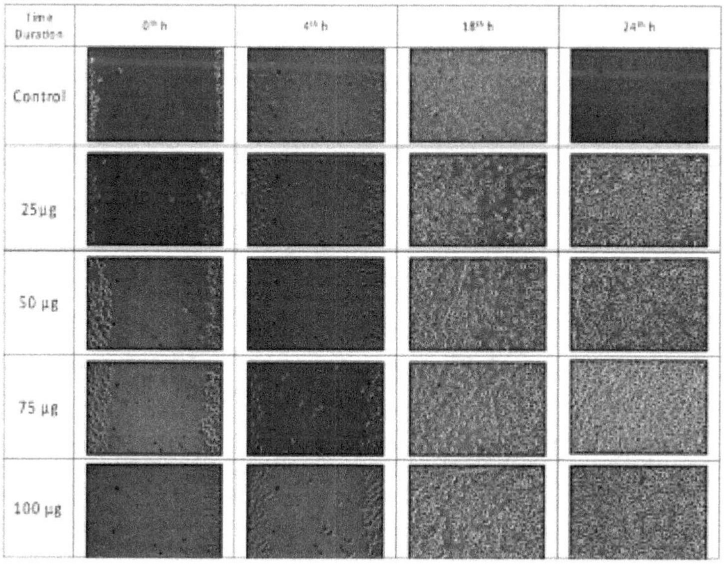

Figure 4.11 Scratch wound assay using L929 fibroblasts incubated with CAP Sponge (25, 50, 75 & µg) at 0, 4, 18, 24 hours

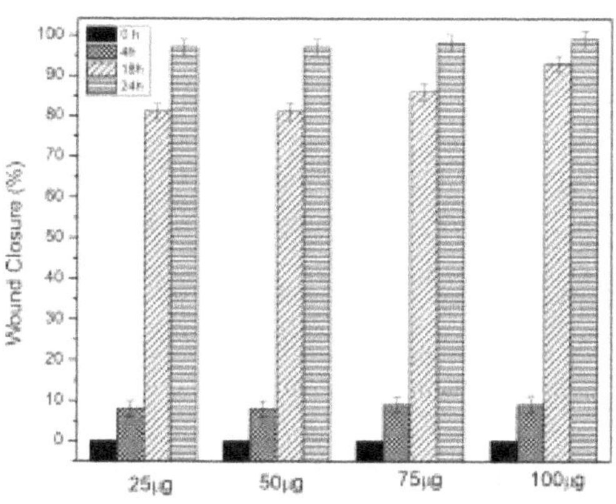

Figure 4.12 Percentage of wound closure at 0, 4, 18, 24 hours. Data represent mean±standard error of the mean(n=3)

4.4 CONCLUSION

Cellulose supported FG 75:25 sponges were successfully fabricated for developing wound dressing for chronic wounds with excessive exudate. The FTIR Spectrum confirms that alum loaded FG 75:25 blends(A) adhered on the cellulose via hydroxyl bonding and it provided the CAP both the absorbing nature of cellulose as well as the surface hydrophobicity and biocompatibility of FG 75:25 blends. This study confirms the absorption capacity of cellulose in all the cellulose supported sponges for absorbing the excess exudates in chronic wounds through various studies like porosity and swelling studies. Cytotoxicity assays are the initial testing to evaluate the biocompatibility of materials and it was confirmed that cellulose supported alum-loaded FG 75:25 blends serve the purpose of absorption of excessive exudates, mild toxicity and slow degradation capacity. Hence, CAP sponge was further considered to study its effect in intrinsic blood coagulation pathway and wound healing activity through scratch assay along the paramylon release kinetics. It was confirmed that CAP sponges show no negative effect on the blood coagulation process and it also has the wound closure activity of 97-99% with burst release at the initial stage upto 12 hours and sustained its release till 48 hours. Thus this sponge can be further studied for knowing its immunological behaviour and *in vivo* in diabetic induced rats and *in vitro* by collecting biopsy from chronic wound patients to confirm its overall behaviour to develop a self-care bandage for chronic wound patients.

CHAPTER 5

EFFICIENCY OF PARAMYLON COATED SILK FIBROIN SPONGE SHEET FOR CHRONIC WOUND HEALING

This chapter deals with the preparation and characterization of paramylon coated on silk fibroin sponge sheet for developing chronic wound dressing.

5.1 INTRODUCTION

The skin plays an integral role in protecting organs against infection, contamination and also forms a barrier against entry of microorganisms from the environment (Grice & Segre 2011). A wound is a disruption of integrity and the continuity of the intact skin, rendering the human body more susceptible to microbial load which may cause delayed wound healing, improper collagen deposition and exudate formation. Wound healing is an unmet therapeutic challenge among medical society (Kadam *et al.* 2019). Natural polymeric-based dressings are of central interest due to their outstanding biocompatibility, biodegradability, low toxicity and non-allergenic nature. Among them, silk fibroin has exceptional characteristics as a wound dressing (Farokhi *et al.* 2018). Silk fibroin derived from Bombyx mori silkworm cocoon is very well-known for its robust mechanical property and has been proven to have prominent characteristics such as good biocompatibility, moisture and oxygen permeability, promoting proliferation of human keratinocytes and human skin fibroblasts (Altman *et al.* 2003) and can be processed to many forms like films

(Kweon *et al.* 2001), hydrogel (Thangavel *et al.* 2017), non-woven mat (Wang *et al.* 2018) and electrospun nano-fibers (Lee *et al.* 2014). Silk fibroin based dressings can also be used as a carriers for delivering drugs, growth factors and bioactive agents to the wound area as it provides appropriate support for complete healing (Farokhi *et al.* 2018).

Silk sponges provide a versatile 3D porous scaffolding material for many applications including wound healing. The Silk sponges can be prepared with the help of two protocols namely aqueous-based or HFIP-based scaffolds (Correia *et al.* 2012). The aqueous-based sponges have excellent interconnectivity between the pores and have the benefit of not requiring an organic solvent. This may be of interest if a bioactive molecule will be incorporated into the silk matrix during processing (Kim *et al.* 2005). The HFIP-based sponges have smoother surfaces along the pores and have higher mechanical strength. Through this processing method, reinforcing agents can be added to the silk matrix to further increase mechanical properties (Nazarov *et al.* 2004). In addition, it has been shown that aqueous-based sponges degrade faster than HFIP-based sponges which enables the user to tailor these scaffolds according to the specifications (Vepari & Kaplan 2007). Also the versatile functionalization nature of SF helps in using it for innovative applications (Nguyen *et al.* 2019). Thus, Silk fibroin is functionalized with a bioactive molecule called paramylon for treating chronic wounds.

Paramylon is a Naturally occurring linear beta-1,3-D-glucan polysaccharide isolated from *Euglena gracilis* Z. Paramylon has been studied as a polysaccharide nanofiber (Shibakami *et al.* 2013), thermoplastic (Shibakami *et al.* 2015) and as a transparent thin film

(Shibakami *et al.* 2015b) which enabled the use of paramylon in the wound care materials.

For functionalization of paramylon on silk fibroin, paramylon was coated over silk fibroin with a help of a fixative called potash alum. Potash Alum is the potassium aluminum sulfate which is used frequently as mordant in the textile industry for fixing the dyes on the fibers specifically for protein and cellulose fibers/fabrics (Haar *et al.* 2013).

5.2 EXPERIMENTAL SECTION

5.2.1 Materials

Bombyx mori silk cocoon was kindly supplied by the Avinaash Silk, Salem. β(1–3)-D-glucan used were derived from *Euglena gracilis* and was purchased from Sigma-Aldrich Chemical Co.(St. Louis, MO.) and used without further treatment. All other chemicals used were analytical grade obtained commercially. L929 cell line (murine fibroblast) was purchased from National Centre for Cell Science (NCCS), Pune.

5.2.2 Methodology

5.2.2.1 Preparation of regenerated silk fibroin solution

Regenerated aqueous silk fibroin solution was prepared by the Rockwood *et al.* (2011). Firstly, the Bombyx mori cocoons were cut into small snippets and treated with the boiling 0.02M Sodium Carbonate solution for 30 minutes under constant stirring for degumming. The degummed fibers were then washed 2 to 3 times with ultrapure water to remove sericin protein completely and its dried overnight under sterile condition. The dried degummed fibers were dissolved in 9.3M Lithium Bromide for 4 hours at 60°C which is then dialysed against sterilized water for three days, changing water every 6 hours in order to remove LiBr. After dialysis, the regenerated silk fibroin solution was kept under -20 °C

5.2.2.2 Preparation of aqueous-based silk fibroin sponge

Aqueous derived silk fibroin sponges were prepared by salt leaching method. NaCl salt was sieved with mesh to obtain particle size distributions between 250-500 µm and added into the 6% aqueous silk fibroin solution at 2:1 (w/v) ratio in a container. The container was covered and left at room temperature. After 24 h, the container was immersed in water to remove salts for 2 days replacing water freshly. Store the sponges in ultrapure water at 4 °C until usage and can be used after drying under vacuum (Correia *et al.* 2012).

5.2.2.3 Preparation of paramylon coated silk fibroin sponge

0.5% paramylon was prepared with 50% formic acid as the solvent and 1% Potash Alum was prepared with ultrapure water. The prepared sponge (SFS) is coated with 0.5% BG is named as BFS and with 1% Potash Alum is named as PFS. The SFS coated with 1% Potash Alum and then with 0.5% BG is named as BPFS. All coatings were done using saturation coating method.

5.2.3 Characterization Studies

5.2.3.1 Morphological studies by scanning electron microscopy

The Surface morphologies and elemental composition of the sponges was examined using Carl Zeiss MA15/EVO 18 SEM coupled with energy-dispersive X-ray (Oxford Instruments NanoAnalysis INCA Energy 250 Microanalysis System (EDX). About 0.5 cm of the dry sponges were cut and sputter coated with gold to enhance the surface conductivity.

5.2.3.2 Fourier transform infrared (FTIR) spectroscopy analysis

Secondary structure and conformational changes were confirmed by FTIR spectroscopy with air as the reference in the spectral range of 4000 - 400 cm-1 at 4 cm-1 spectral resolution and 32 scans using JASCO International Co./Japan Fourier transform infrared Spectrometer Model FTIR-6300 instrument.

5.2.3.3 Surface wettability testing - contact angle measurement

According to Mates *et al.* 2014, the procedure followed in the section 3.2.3.3 for nonwoven mat is followed here to measure the wettability of all the SF based sponges.

5.2.3.4 Porosity measurement by solvent replacement method

To find the porosity of the silk fibroin based sponges, the steps or procedures followed in section 4.2.3.3 was followed and the porosity was calculated from the below formula (Nanda *et al.* 2013)

$$\text{Porosity} = [(M_2 - M_1)/\rho V] \times 100 \quad (5.1)$$

5.2.3.5 Swelling capacity by gravimetric method

The determine the swelling capacity of the Silk fibroin based sponges, the procedure involved in section 4.2.3.6 was followed and then, the swelling index (SI) was calculated with the below formula (Zeimaran *et al.* 2015)

$$SI = \{(W_s - W_d) / W_d\} \times 100 \quad (5.2)$$

5.2.3.6 *In vitro* enzyme degradation studies

In order to find the enzyme degradation property of silk fibroin based sponges, the procedure or steps followed in 2.2.1.4.A was followed and the percentage of the weight remained after degradation was analysed using the formula below (Nwe *et al.* 2009).

% Weight remained = $(W_t / W_0) \times 100$ (5.3)

5.2.3.7 Blood compatibility testing – *in vitro* haemolytic activity

Haemolysis activity assay was conducted to evaluate the blood compatibility of the sponges. For this testing, the procedures followed in section 4.2.3.6 were followed and the resulted supernatant was taken and its absorbance at 545 nm was recorded. The % Haemolysis was calculated using the following equation (Kim *et al.* 2014),

Haemolysis (%) = $[(A_{\text{test sample}} - A_{(-)\text{ control}}) / (A_{(+)\text{ control}} - A_{(-)\text{ control}})] \times 100$ (5.4)

5.2.3.8 Cytotoxicity Testing – *In vitro* Cell Viability Assay

Viability of the cells was performed using MTT assay. For this assay, the steps involved in section 3.2.3.8 was repeated on silk based sponges against L929 cells and the resulting blue precipitate was solubilised in Dimethyl sulfoxide (DMSO) and the absorbance was recorded at 570 nm wavelength using a standard plate reader (Wiegand *et al.* 2019). The percentage cell viability (CV) was calculated using the formula below:

CV(%) = (Absorbance of test sample / Absorbance of control) x 100 (5.5)

5.2.3.9 Wound healing Studies

To confirm the biological activity of the Paramylon functionalized SF sponge, either BFS or BPFS can be considered but on comparing BPFS was less toxic and has moderate water absorption capacity and thus BPFS was further considered to study the wound healing activity as it showed significant results in swelling ability, Hemocompatibility and cell viability studies.

A. Dynamic Blood Coagulation Assay

Dynamic blood coagulation assay measures the release of haemoglobin from residual erythrocytes that remained free from entrapment during clot formation. The Assays were performed by recalcification of the anticoagulated whole blood. Briefly, BPFS Sponge sheet were placed at the bottom of 100 ml beaker and then prewarmed in a water bath at 37 °C for 5 min. Eventually, the steps followed in section 2.2.1.5.A was followed and at the designated termination point, the beaker was removed from the water bath and shaken for 10 min followed by addition of 50 ml distilled water to lyse the free erythrocytes. The absorbance of the supernatants at 540 nm was determined. The absorbance signal derived from the proportion of clot-free erythrocytes are inversely proportional to the size of the clots. The absorbance-time curve was constructed (Liu *et al.* 2014).

B. *In vitro* Paramylon Release Kinetic Assay

This Assay was performed to investigate the release kinetics (burst release vs. sustained release) of paramylon blended with silk fibroin. In order to evaluate the drug release kinetics, BPFS sponge sheet were cut into samples of average weight of 5 mg and then the procedures followed for films in section 2.2.1.5.C is followed here also and UV spectroscopy was used for detecting the paramylon release profile at 208 nm. At the beginning of the release test, a linear calibration curve of standard paramylon solution from 5 to 20 μg/ml was created. Paramylon release behaviour in the BPFS sponge sheet were analyzed according to first order model (Cam *et al.* 2020).

C. *In vitro* wound scratch assay

To monitor the cell proliferation via wound scratch assay, the procedures followed in section 2.2.1.5.C is followed here in BPFS sponge and the cell proliferation was monitored at different time points: 0 h, 4 h, 18 h, 24 h and images of the migrated cells were taken at all different time points using digital camera (Nikon, Tokoyo, Japan) connected to the inverted phase contrast microscope (Radical instruments, India).

5.3 RESULT AND DISCUSSIONS

5.3.1 Characteristics of Silk fibroin based sponges

All the silk sponges that were coated with paramylon or/and potash alum via saturation coating method appeared to be porous, spongy and white in colour. The thickness of the sponges were around 0.25 cm and the architecture of all the sponges was characterized by SEM (Figure 5.1). The SEM topography showed the porous nature of all the silk-based sponges and also shows the irregular pattern in PFS and BFS due to the shrinkage and rupturing of sponges.

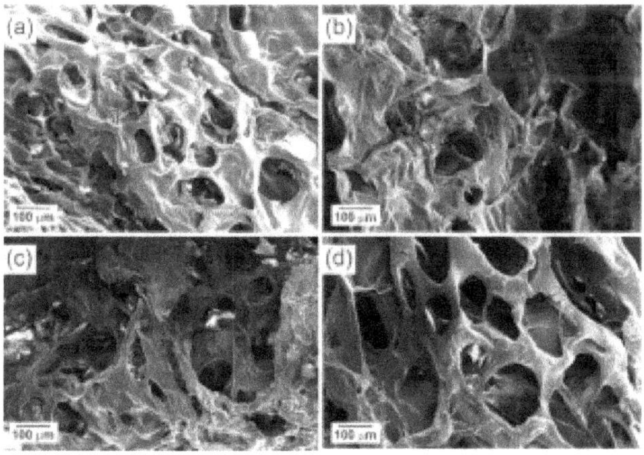

Figure 5.1 Surface morphology of (a) SFS, (b) PFS (c) BFS (d) BPFS

5.3.2 Molecular Conformation Studies by FTIR Analysis

IR spectroscopy is commonly used to investigate the conformation of SF and BG since the IR spectrum represents typical absorption bands sensitive to the molecular conformation (Kweon *et al.* 2001). Among all the amide bands of protein, amide I, II and III are generally selected to differentiate protein structure which were in the range of 1700-1600 cm-1, 1600-1500 cm-1 and 1300-1200 cm-1 respectively (Arthe *et al.* 2020). The bands at 1622 cm-1 and 1524 cm-1 corresponds to the amide I and amide II linkage of the Silk fibroin protein with β-sheet conformation. Another band at 1230 cm-1 corresponds to the amide III linkage with random coil configuration (Ha *et al.* 2005). Due to the least concentration of paramylon, the bands corresponding to this polysaccharide is not more predominant but the band at 1060 cm-1 broadens slightly due to presence of sugar moiety in the Paramylon (BG) (Synytsya &

Novak 2014). The sharp peak at 3283 cm-1 corresponds to the N-H stretching in Silk Fibroin. The intensity of the peak is reduced in the BPFS due to the predominant coating of paramylon on Silk fibroin with the help of Alum.

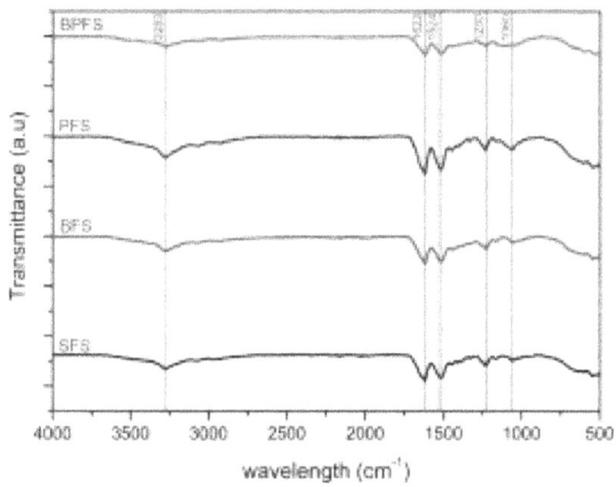

Figure 5.2 FTIR spectrum of Functionalized and Non-functionalized Silk fibroin sponge

5.3.3 Surface Wettability Testing

The surface wettability of the silk fibroin-based sponges functionalized with paramylon was analysed by sessile drop method. An ideal dressing should be able to absorb the wound exudate and maintain moisture and these criteria are influenced by the surface wettability and hydrophilicity behaviour of the dressing (Samadian *et al.* 2020). From the Figure 5.3, it is clear that all the paramylon coated Silk fibroin sponges showed high to moderate wettability as according to Mitta (2003) the contact angle from 0°to 60° shows high wettability and from 60° to 90° shows moderate wettability.

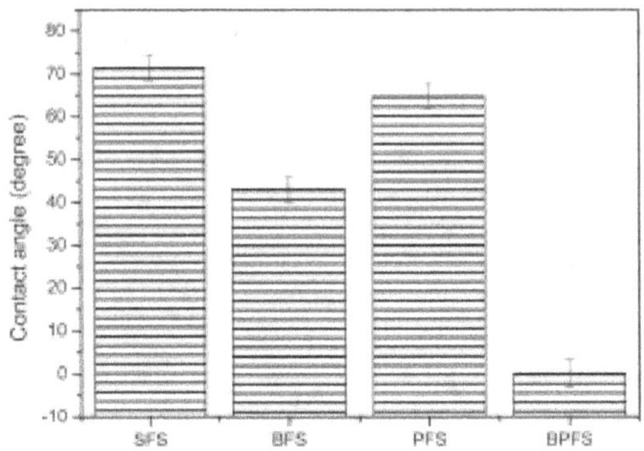

Figure 5.3 Water Contact angle of Functionalized and Non-functionalized Silk fibroin sponge. Data represents the mean±standard error (n=3)

5.3.4 Porosity Measurement

In general, a desirable porosity and the morphology of wound dressing are essential to maintain moisture and to accelerate the wound healing (Negut 2020). The solvent replacement method was used to measure the porosity of the sponges. The result showed that the porosity of SFS, BFS, PFS and BPFS were 92.857±2.9, 82.33±2.3, 57.49±2.45 and 61.99±2.35

respectively. The Figure 5.4 reveals that the Silk fibroin sponge (SFS) shows higher porosity which was decreased in all other sponges due to the water expansion, shrinkage and rupturing of sponges. Chong *et al.* (2007) concluded that the porosity in the range of 60-90% is preferred for tissue engineering applications. In another study, Shan *et al.* (2015) reported the porosity range of around 87% for nanofibrous dressing. Thus, the porosity of the fabricated sponges was in the acceptable range.

99

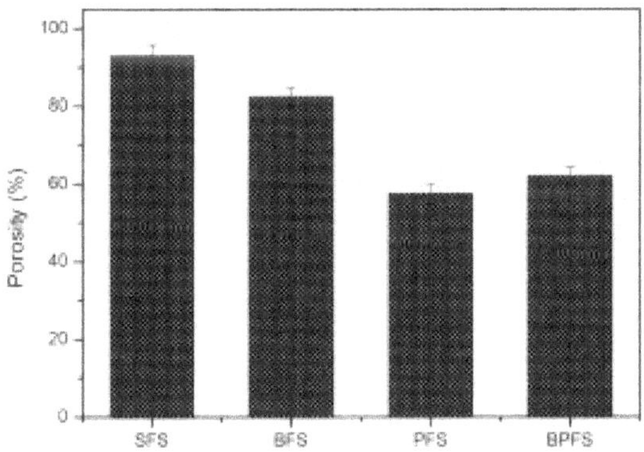

Figure 5.4 Porosity percentage of Functionalized and Non-functionalized Silk fibroin sponge. Data represents the mean±standard error (n=3)

5.3.5 Swelling Capacity Studies

As far as surface characteristics are concerned, the hydrophilicity of the materials has been shown to be an important factor for wound dressings. The Swelling capacity measurement is a gravimetric test that determines the maximum amount absorbed and retained by the dressing as a percentage (Ngadaonye *et al.* 2013). Penetration of solvent leads to swelling of the sponges due to the diffusion of water molecules. Figure 5.5 showed the fluid holding capacity of the silk fibroin-based sponges. Due to the porous nature of silk fibroin sponge, the water molecules diffuse into the SFS very well compared to BFS, PFS and BPFS. Meanwhile, the difference in the swelling index in water and PBS is caused by a higher ion concentration of PBS that leads to low osmotic pressure resulting in a lower absorption capacity (Kusumastuti *et al.* 2019). It was showed that the swelling ratio of PFS and BPFS was higher in PBS when compared to water, which was probably due to the ionic nature of the sponge as it contains the potash alum coating on it.

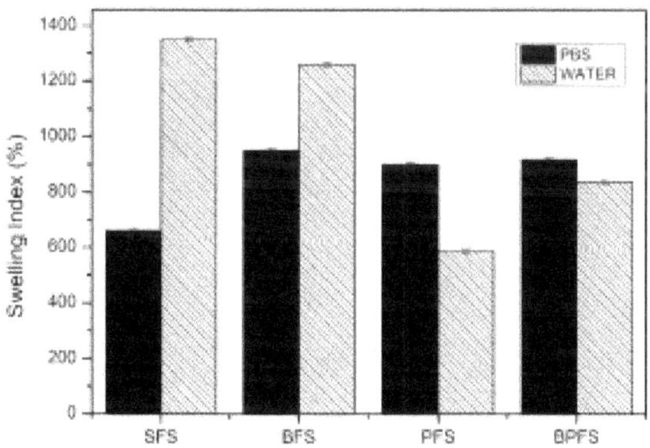

Figure 5.5 Swelling ratio of Functionalized and Non-functionalized Silk fibroin sponge. Data represents the mean±standard error (n=3)

5.3.6 *In vitro* Enzyme Degradation Studies

For wound dressing purpose, it is important that the dressing should be stable and should not degrade too fast, as degradation may lead to the release of absorbed exudates, reduces hydration and mechanical stiffness. Also, nondegradable materials can also cause problem in disposal and hence the ideal degradation rate and breakdown matches for wound healing application (Veld *et al.* 2020). The biodegradation of the wound dressing sponge was studied by lysozyme hydrolysis. The residual weight of the sponges after immersion were inversely proportional to the degradation rate (Huang *et al.* 2010). From Figure 5.6, it has been observed that all the sponges maintained nearly 75% of its original weight for about 3 weeks.

101

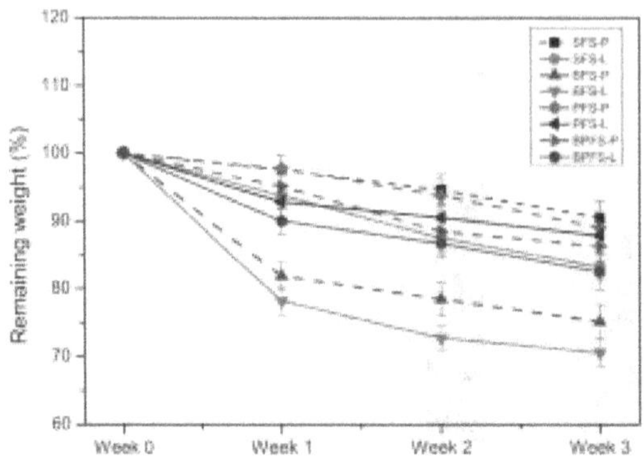

Figure 5.6 Weight remaining percentage of Functionalized and Non- functionalized Silk fibroin sponge in enzymatic and non- enzymatic degradation environment. Data represent mean±standard error of the mean(n=3)

5.3.7 Blood Compatibility Testing – *In vitro* Haemolytic Activity Assay

Haemolysis analysis has been done to evaluate the adverse effect of the sponges on the blood cells. During haemolysis, red blood cells are disrupted and the free Hb can be detected in the surrounding environment and influence the wound healing process by causing Nitric oxide inactivation and disturb fibroblast proliferation and collagen remodelling (Qu *et al.* 2018). From Figure 5.7, it was observed that all the sponges obtained values closer to that of the negative control which confirms that these sponges do not show any traces of disintegration of blood cells and shows good blood compatibility.

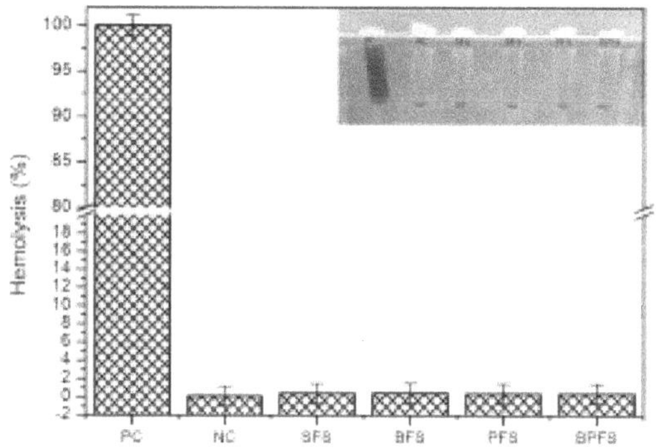

Figure 5.7 Percentage of Haemolytic activity in Functionalized and Non-functionalized Silk fibroin sponge. Data represent mean±standard error of the mean(n=3)

5.3.8 Cytotoxicity Testing – *In vitro* Cell Viability Assay

The MTT reduction assay was performed to investigate the cytotoxicity of the sponges based on the effect of sponge with on the viability of L929 cells after the incubation of 24 hours. Cytotoxicity is inversely proportional to cell viability rated as severe (<30%), moderate (30-60%), mild (60-90%) or non-cytotoxic (>90%) based on the activity relative to values obtained for the controls (Dioguardi *et al.* 2015). From the Figure 5.8 it was observed that all that sponges show moderate cytotoxicity which may be due to the porosity of the sponges against the L929 cells. As according to Lee *et al.* (2020) in L929 cell viability assay, the relationship between porosity and cell viability showed a similar trend with MTS assay, but the cell viability started to decrease as the porosity went too high (70%). When the porosity goes too high, it may cause a decrease in cell viability (Lee *et al.* 2020)

103

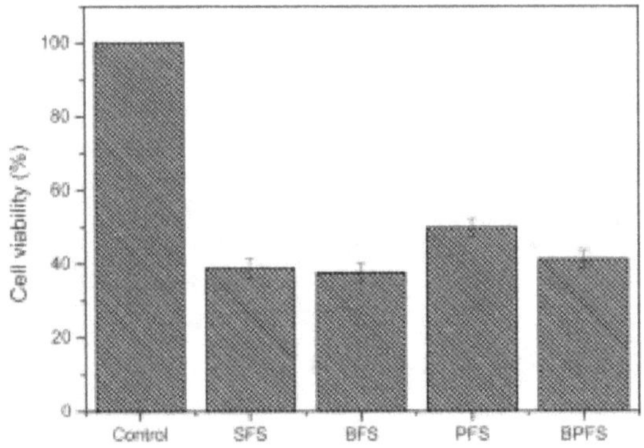

Figure 5.8 Effect of Functionalized and Non-functionalized Silk fibroin sponge on L929 cell viability. MTT assay of L929 cells for 24

h. hours. Data represent mean±standard error of the mean(n=3)

5.3.9 Dynamic Blood Coagulation Assay

The dynamic blood clotting time is an *in vitro* test that measures the degree of activation of intrinsic coagulation factors when surfaces interact with blood. Activation is measured by the absorbance-time curve for release of haemoglobin from erythrocytes that avoid being entrapped during clot formation (Liu *et al.* 2014). The curve (Figure 5.9) implies that the sponge shows similar pattern as on control and hence it does not affect the coagulation pathways. Despite this, the clot sizes appear to be bit higher in BPFS sponge sheet, since the magnitude of absorbance is inversely related to clot size.

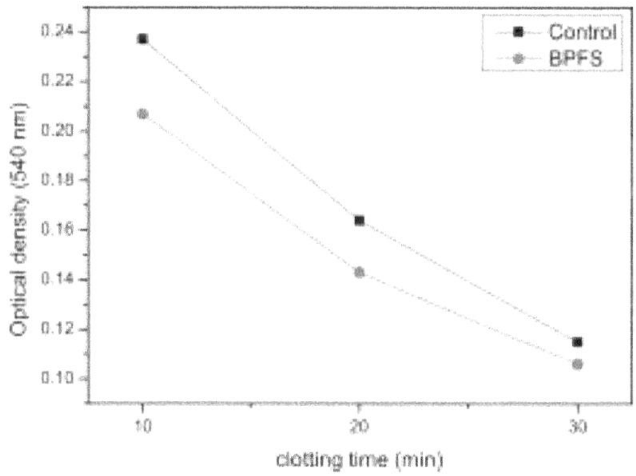

Figure 5.9 Intrinsic coagulation factors activation measurement curve.

Data represent mean±standard error of the mean(n=3)

5.3.10 *In vitro* Paramylon Release Kinetic Assay

In vitro drug release kinetic assays for BPFS sponge sheet were performed according to the first-order model. Primarily, the UV spectra were obtained with the concentration range of paramylon from 5 to 20 μg/ml and a linear standard calibration curve was drawn from paramylon absorption values (R2 = 0.99127) obtained for the quantitative determination from UV absorbance measured at 208 nm. The *in vitro* release studies were performed over a period of 48 hours for BPFS sponge sheet in order to determine the release kinetics of blended bioactive paramylon. The sponge showed a burst release in approximately 8 hours and maintained a sustained release after 12 hours with low concentration of BG (20%) (Figure 5.10). These results are inconsistent with previous studies in which the burst release reaches high concentration of bioactive molecule and a sustained release afterwards may help in wound closure but here due to the low concentration release of BG least migration is showed in the wound scratch assay (Cam *et al.* 2020).

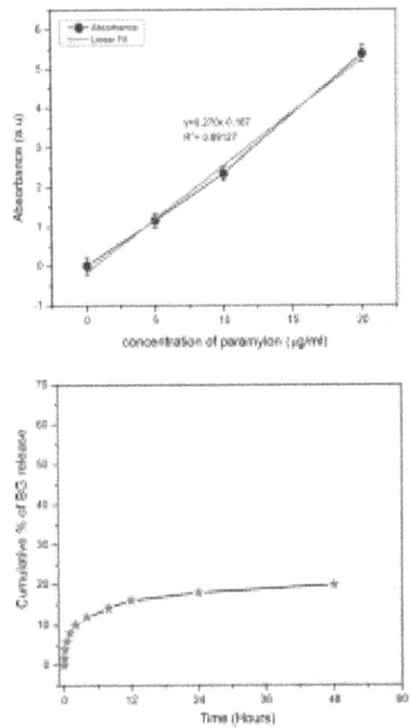

Figure 5.10 *In vitro* drug release profile of Functionalized Silk fibroin sponge: (a) Paramylon calibration curve (b) Paramylon release profile of BPFS sponge sheet according to first-order model. Data represent mean±standard error of the mean(n=3)

5.3.11 *In vitro* Wound scratch assay

Cell migration is an essential process in many physiological and pathological events such as wound healing, embryonic development and cancer metastasis. The basic procedure to study cell migration involves (i) creation of a cell monolayer, (ii) the production of a pseudo-wound in the confluent layer of cells, (iii) the capture of images at different time intervals until wound closure is reached, and (iv) the analysis of the image sequence in order to quantify the migration of the chosen cells (Cappiello *et al.* 2018). The *in vitro* scratch assay is an easy and economic method to mimic the cell migration during wound healing *in vivo*. L929 cells were incubated in presence and absence of the BPFS sponge sheet and images were captured at 0, 4, 18 and 24 h at various concentrations of BPFS sponge sheet (25, 50, 75 and 100 μg). From the Figure 5.11, it is clearly observed that the cell migration started 4 hours after incubation in almost all the concentration and it was only 37-49 % after 24 hrs. Therefore, the wound healing capacity of BPFS sponge sheet on L929 Fibroblast cells is less compared to that of other samples and silk fibroin-based sponge should be further optimised to produce the sponge with less toxicity and good cell migration capability.

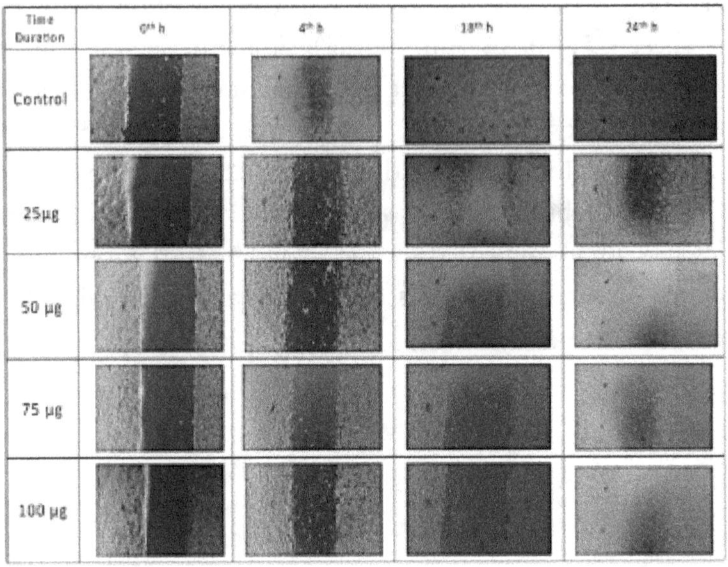

Figure 5.11 Scratch wound assay using L929 fibroblasts incubated with BPFS sponge sheet (25, 50, 75 & µg) at 0, 4, 18, 24 hours

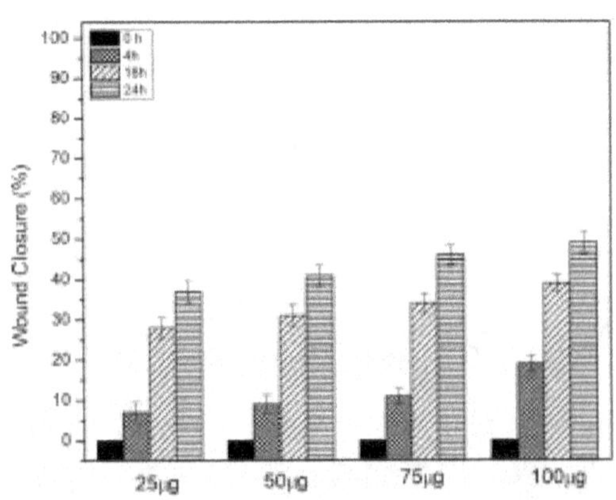

Figure 5.12 **Percentage of wound closure at 0, 4, 18, 24 hours. Data represent mean±standard error of the mean(n=3)**

5.4 CONCLUSION

Paramylon coated Silk fibroin sponge were successfully fabricated for developing wound dressing for chronic wounds. The FTIR Spectrum confirms the presence of amide I, II and III bands. From the porosity, wettability, swelling capacity and degradation studies, the BPFS sponges showed significant property for developing wound dressing material. But it also showed mild toxicity as it showed 58.8% of cytotoxicity which was predicted due the pore size of the sponge and to confirm this further this BPFS sponge was further consider to study its effect on the intrinsic coagulation pathway and to study the wound healing activity along with the release percentage of the paramylon through first order kinetics. From all the above three analysis, the BPFS sponge does not show any negative effect on blood coagulation but it showed lesser paramylon release of only 20% (0.5% paramylon) and also showed lesser cell migration rate of about 37-49% in the wound scratch assay. Thus Silk fibroin based sponge preparation has to be further optimized develop a significant porosity and produce a non-toxic sponge with good cell migration property to develop a wound dressing material for chronic wound patients.

CHAPTER 6

SUMMARY AND SUGGESTIONS FOR FUTURE WORK

This chapter summaries about the overall findings of the present work. It also deals with the overall inference and the suggestions for future work.

6.1 SUMMARY OF THE WORK

Chronic skin wounds are one of the dangerous issues that are achieving epidemic proportions that are predicted to affect 20-60 million people around the world by 2026. The endless rising of obesity, diabetes, aging population and changes in lifestyle prolongs the frequency of the chronic wounds. These wounds are more than just a cost burden, as they create immobility and reduces the ability to perform daily tasks, loss of participation in the work and worthlessness of life. The impact of this non- healing wounds on mortality has even been similar to or worse than that of few common cancers (da silva *et al.* 2019). Some common features of chronic wounds include prolonged or excessive inflammation, persistent infections, formation of microbial biofilms and inability of cells to respond to reparative stimuli. The prolonged inflammation leads to elevated levels of proteases which in turn degrades the ECM and growth factors responsible for wound healing. The Extracellular matrix play a significant regulatory role in the biological behaviour of epidermal cells as it helps in migrating them from the wound edges by being in contact with the help of fibronectin, fibrin and collagen. Typically, in chronic wounds the epithelial cells proliferate and

109

bunch up on the wound edges but fail to migrate. Lack of a normal ECM at the wound edges is now recognized to be the real problem (Donaldson & Mahan 1983). The endowment of therapeutic dressing to positively regulate MMPs and inspire the function of ECM helps in fibroblast migration at wound site. Recently, natural polymers have highly attracted the scientifical community interest (Chantre *et al.* 2019). By knowing the biocompatibility and biodegradability property of the natural polymers, highest level of biomimicry can be attained by replicating the biological and physico- chemical features of the native ECM. The increasing curiosity in the use of either protein-based or polysaccharide-derived dressings provides astonishing and growing approach of giving back what we borrowed from Nature (Suarato *et al.* 2018). Thus, Silk fibroin protein and paramylon polysaccharide has been used in different forms to develop a wound dressing material for accelerate the chronic wound healing.

The Chapter 2 reviews about the silk fibroin and paramylon along with its importance in the wound healing process. This Chapter also deals with the preparation of Silk fibroin and paramylon blended films along with the *in vitro* compatibility testing of the films for developing further forms of these blends to prepare a bioactive wound dressing. Thus, here in this chapter, Silk fibroin and paramylon blends were successfully prepared in the ratio 0:100, 25:75, 50:50, 75:25, 100:0 respectively. Various biological characterizations like *in vitro* enzyme degradation studies, blood compatibility studies, cytotoxicity studies were studied. All the films showed good blood and biocompatibility and also retained about 80% of their total weight during both enzymatic and non-enzymatic degradation for the period of 4 weeks. For further development, one blend film with significant property i.e. FG 75:25 film was considered for knowing the wound healing property of the blend films and hence the dynamic blood coagulation assay, paramylon release and scratch assay was carried out. It was confirmed that the FG 75:25

film does not show any effect against the intrinsic coagulation pathway and can help in Haemostasis process. Also, the wound scratch assay resulted in 91-98% of wound closure. This chapter concludes that FG 75:25 film showed good wound healing activity and hence this can be further processed and studied *in vivo* to develop a self-care bandage for chronic wound healing.

In Chapter 3, Silk fibroin based nonwoven mats functionalized with 0.25%, 0.5% and 1% of paramylon were fabricated as nonwoven processing is very easy and economical. The Characteristics of all the functionalized and non-functionalized mat was analysed by surface morphological studies, molecular confirmation through FTIR, wettability through contact angle measurement, porosity, swelling and *in vitro* enzyme degradation studies along with blood compatibility and Biocompatibility. On comparing with all the nonwoven mat, SBM25 showed good wettability, porosity, swelling capability and slower degradation rate of upto 80% for 2 weeks. Also SBM25 showed good blood compatibility and nontoxic too. Hence, SBM25 was further considered to study the wound healing property of the nonwoven mat, which showed that the SBM25 is good for the intrinsic blood coagulation pathway and showed 92-99% of wound closure in the scratch assay with the initial burst release of paramylon through the first order kinetics. Thus, this formulation can be further studied *in vivo* with streptozotocin induced rats to develop a self-care bandage.

The Chapter 4 deals with the development of wound dressing material for heavily exudating chronic wound. So, highly absorbing cellulose based sponges were used. FG 75:25 blends loaded with Alum were coated on the cellulose sponge was fabricated to study the characteristics for developing a highly absorbing self-care bandage for chronic wound patients. For the study, other combinations like with and without alum loaded FG 75:25 blends were also prepared which was coated on cellulose sponge and also developed as films without coating for comparing. All the sponges showed the porous nature, high absorption capacity and good blood compatibility. The sponges showed mild to moderate toxicity due to the presence of Magnesium chloride that was used for softening the commercial cellulose sponge which was confirmed through the elemental analysis. To further study the wound healing property, the CAP was considered as it showed moderate toxicity due to the predominant coating of FG 75:25 blends. The Paramylon release was studied in CAP and was used to study the wound closure through scratch assay along with its effect in intrinsic blood coagulation. It was confirmed that the CAP does not show any adverse effect on blood coagulation but helped in making the clot size bigger to reduce the blood loss due to the presence of Alum in it. Also, it showed a burst release initially that wound have helped in the 97-99% of wound closure with 25, 50 ,75 and 100μg of CAP.

The last chapter 5 deals with the fabrication of Silk fibroin sponge and functionalization of SF sponge with paramylon with/without potash alum through saturation coating. Like previous chapters, surface morphology, wettability, porosity, swelling, degradation capacity, blood compatibility and biocompatibility was studied. The paramylon coated SF sponge with alum showed good wettability, moderate porosity, swelling capacity and slow degradation capacity up to week 3. They also showed good blood compatibility and moderate toxicity due to its microporous nature according to the Lee *et al.* 2020 and also with the pattern of porosity percentage with other silk-based sponges. Due to its good physical nature, BPFS was further studied to confirm its wound healing property. Unfortunately, it showed only 37-49% of wound closure in the wound scratch assay and least paramylon release percentage although it doesn't show any effect on blood coagulation. Thus, this silk based sponge has to be further studied and optimized to develop wound dressing sponge with no toxicity and good cell migration capacity.

6.2 SUGGESTIONS FOR FUTURE WORK

The silk fibroin protein and paramylon polysaccharide complex has been developed in different forms. Although all the forms except BPFS showed good wound healing property, still they have to studied further based on the release of paramylon. According to literatures, all beta glucan are immunomodulating and thus it has to be studied further by knowing its effect on immune cells and growth factors. Also this wound healing capacity has to be tested *in vivo* with the streptozotocin induced rats and also *in vitro* by collecting the chronic wound biopsy from patients along with the *in vitro* studies to confirm its capacity in angiogenesis and epithelialization process.

www.ingramcontent.com/pod-product-compliance
Lightning Source LLC
LaVergne TN
LVHW052317200926
848481LV00002B/588